IEE Telecommunications Series 11

Series Editors: Prof. J.E. Flood
C.J. Hughes

Principles and practice of multi-frequency telegraphy

Previous volumes in this series

Volume 1	Telecommunications networks J.E. Flood (Editor)
Volume 2	Principles of telecommunication-traffic engineering D. Bear
Volume 3	Programming electronic switching systems M.T. Hills and S. Kano
Volume 4	Digital transmission systems P. Bylanski and D.G.W. Ingram
Volume 5	Angle modulation: the theory of system assessment J.H. Roberts
Volume 6	Signalling in telecommunications networks S. Welch
Volume 7	Elements of telecommunications economics S.C. Littlechild
Volume 8	Software design for electronic switching systems S. Takemura, H. Kawashima, N. Nakajima Edited by M.T. Hills
Volume 9	Phase noise in signal sources W.P. Robins
Volume 10	Local telecommunications J.M. Griffiths (Editor)

Principles and practice of multi-frequency telegraphy

J.D. Ralphs

Peter Peregrinus Ltd on behalf of The Institution of Electrical Engineers

Published by: Peter Peregrinus Ltd., London, UK.

While the author and the publishers believe that the information and guidance given in this work is correct, all parties must rely upon their own skill and judgment when making use of it. Neither the author nor the publishers assume any liability to anyone for any loss or damage caused by any error or omission in the work, whether such error or omission is the result of negligence or any other cause. Any and all such liability is disclaimed.

British Library Cataloguing in Publication Data

Ralphs, J.D.
Principles and practice of multi-frequency telegraphy. — (IEE telecommunications series; II)
1. Telegraph, Wireless
I. Title II. Series
621.384'2 TK5741

ISBN 0-86341-022-7

Printed in England by Short Run Press Ltd., Exeter

Errata

Reverse of title page should read:

Published by: Peter Peregrinus Ltd., London, UK.

Contents

Preface

In the first chapter of this book I have explained the technical background and origins of multi-frequency shift keying (MFSK), but that does not tell the whole story. Over the last 25 years or so, the study of MFSK principles and the design of equipment based on those principles has occupied some 30–50% of my working time and has become something of a personal project. This may be apparent in the text of the book, but would in any case be made evident by the use of such phrases as 'in the author's opinion'. My concern is that this obvious individual involvement may give the impression that I have attempted to conceal or belittle the contribution of others, which would be unforgivable on my part. Since the text of a technical monograph is not the most appropriate place to discuss such matters I would like to clarify the situation here.

The Communication Engineering Department of the (British) Foreign and Commonwealth Office (and its predecessor the Diplomatic Wireless Service) is a very practical and down-to-earth organisation which carries out a difficult and demanding engineering operation in installing and maintaining a complex communication network to very high standards. As such, its duties do not normally require it to carry out research in any form, and its engineering staff adopt, of necessity, a pragmatic approach. It is not surprising, therefore, that the early experiments on telegraphy systems carried out in 1949–53 by the then Chief Engineer, Harold Kilner Robin, were entirely empirical. Even had adequate analytic methods been available to him at the time it is doubtful whether he would have employed them, having a deep mistrust of the orthodox approach and an amazing capacity for highly original and complex thinking in purely physical terms. His ideas were ably translated into hardware and experimentally investigated by two transmitter engineers, Tom Murray and Ralph Muggeridge.

When I was asked to lead the design team for the first Piccolo equipment in 1957 I had no previous experience in h.f. telegraphy and was quite happy to accept a task defined by a general description of the overall functions required, plus a detailed study of the existing experimental units. My brief, as a designer of electronic equipment, was simply to convert those principles into a practical telegraphy modem for use in the field, using a new circuit technology and new parameters based on a simple rule-of-thumb relationship.

In the years that followed the introduction of the Piccolo Mark 1 we discovered that when Ralph Waldo Emerson said 'If a man makes a better mouse-trap than his neighbour, the world will make a beaten path to his door', he was being either humorously ironic or incredibly naive. The communication industry was simply not interested in what we saw as a significant advance in telegraphy techniques. In investigating for myself the reasons for this, I became interested in communication theory and also in the very pragmatic world of h.f. telegraphy. I must place on record here my gratitude to a co-operative management (and particularly to the unfailing support and encouragement of my friend, mentor and superior officer for some 20 years, Donald Bayley) in that this study (and the consequent extra-mural activities) was allowed to fill any time not specifically required for higher priority tasks, and accepted to the extent that its conclusions were incorporated in the later models of Piccolo.

To most of my engineering colleagues I was a species of 'boffin'* and quite harmless if left alone. They had their work to do in producing and testing the resulting hardware and integrating it into a complex communication system, which had then to be installed, maintained and operated. They had neither the time nor the inclination to get involved in abstract theory or to study the work of other organisations. However, they were always willing to advise on practical aspects of their profession and to supply the technical feedback I needed from measurements and observations on working radio links. Without that assistance, and the wealth of experience from which it originated, this book would lose most of any value it may have as a guide to the design of practical communication systems, and it is important to recognise that the 'author's opinion' relies very heavily on that collective expertise.

The overall result was that some 20 years later, as the time of my retirement approached, a disturbing situation became apparent. The first of the Piccolo Mark 6 units had been installed and shown to fulfil the requirements of the FCO. However, the fundamental knowledge embodied in that design was (apart from two or three published papers) scattered with other work in a number of largely indecipherable notebooks. Any future extension, adaptation or improvement of Piccolo equipments would be difficult and would probably involve considerable reworking of problems already solved. I was asked to produce a clear and complete exposition of MFSK principles for in-house use, and in view of my own

---

*Boffin: a word coined by Royal Air Force aircrew during the Second World War to describe a 'back-room' scientist working on secret equipment.

interest in the wider applications of MFSK I asked for permission to publish this, together with such clarification and additional matter as would be useful in a more general context.

This book is the result, and many of its faults may be traced to its origins. It has been written by a hybrid scientist–engineer with no specialised training in the highly complex field it purports to discuss. The work described was largely undisciplined in the academic sense, since it was open-ended, pragmatic, not subject to criticism, review or guidance by better qualified authorities, and of necessity incomplete. The sole criterion applied by my superiors and colleagues was that the resulting telegraphy equipments should be effective, and in this respect, thank Heaven, I seem to have succeeded. In view of the high standards they demand and the technical difficulties they have to face, perhaps this fact alone is sufficient justification for my belief that the book is worth publishing.

It is evident, therefore, that I must take full and complete responsibility, not only for the ideas advanced and the opinions expressed, but also for all the mathematical relationships other than those specifically attributed. I apologise in advance for any errors, ambiguities or misunderstandings and would be grateful if these could be brought to my attention.

Finally, I must acknowledge with deep gratitude the work of my immediate colleagues in the Development Section, whose dedication, expertise and cheerful friendliness have made my work so enjoyable over many years. Unfortunately there are too many to name individually, but my overwhelming debt is to Dick Neale, who joined the section 'for a few weeks' in 1959 and stayed to be my deputy, collaborator and unfailing support for some 23 years. Without his work, and theirs, this book could not have been written because there would have been nothing to write about.

J. D. Ralphs
March 1983

Acknowledgements

I am grateful to the Director of Communications of the British Foreign and Commonwealth Office, and to the Controller of Her Majesty's Stationery Office for permission to publish this book, copyright of which is retained by HMSO.

My thanks are also due to the following organisations:

British Telecom, for permission to quote from Ref 6.3 and to reproduce Fig 7.6 from information in that reference.

SHAPE Technical Centre, The Hague, for permission to quote in Para 7.5 from reference 1.9.

Racal Communications Ltd, Bracknell England, for permission to refer in Para 7.5 to the unpublished results of trials in which they assisted.

The communication equipments referred to as the Racal LA 1117 and the Racal LA 2028 were designed and manufactured by that company to specifications issued by the Communication Engineering Department of the Foreign and Commonwealth Office, and were based heavily on design information provided by the Department. This information was also purchased (non-exclusively) by the company for commercial use.

J. D. Ralphs

Chapter 1

Introduction and History

1.1 Introduction

With the inception of satellite communication and the expansion of the submarine cable network, the role of h.f. communication has undergone a drastic change. The removal of the dominating effects of international trunk telegraphy and telephony networks has allowed a greater emphasis on the development and use of smaller, self-contained networks, short-range communication across difficult terrain, emergency links, medium- and short-range telemetry and similar applications. This change of emphasis does not seem to have brought about the detailed re-evaluation of modulation methods which might have been expected, and most of the research and development in this field over the last 20 years has concentrated on the application of ancillary techniques (such as error coding and adaptive systems) to well-established modulation principles, rather than the investigation of more complex basic modulation and detection methods made available by advances in technology and communication theory. This is regrettable, since there is evidence that the use of more effective modulation techniques (together with a re-assessment of the effects of the ionosphere), could discourage the present tendency towards the use of unnecessarily high transmitter powers, and this, in conjunction with the removal of high density trunk links, could considerably relieve the spectrum overcrowding and consequent interference which is one of the major drawbacks of the medium.

One potentially useful type of modulation, completely neglected in the past because of its complexity (and consequent high cost and unreliability with older technology), is multi-frequency shift keying (MFSK).

As far as the Western world is concerned, the use of MFSK techniques has, with minor exceptions, been limited to a single organisation, the UK Foreign and

Commonwealth Office (FCO). It is a rare enough event for a new principle to appear in such a long-established field of technology as h.f. telegraphy and arguably unique for that principle to be empirically investigated and then theoretically justified, for a practical field equipment to be designed and developed, then produced in medium quantities and installed on a world-wide network to be operated successfully for some 15 or more years, and all within the confines of one relatively small communication authority. This, however, summarises the history of the 'Piccolo' equipments, which have, since the mid-1960s, provided the basis for the h.f. telegraphy network between the British Government in London and many British Embassies overseas.

Within that organisation the effectiveness of the technique is completely accepted. It has been demonstrated on countless occasions that, with few exceptions, if a signal is at all audible then communication is possible with adequate accuracy for most purposes, and it is a routine occurrence for signals to be successfully worked despite their being totally undetectable by the ear of an experienced Morse operator. It is relevant to comment that there are few engineers with experience limited to more conventional techniques in h.f. telegraphy who would undertake to design a world-wide point-to-point network exchanging high priority telegrams over routes from 1 500 to 12 000 km long, to small outstations equipped with 8 m whip or monopole aerials, a transmitter of less than 500 W, and invariably situated in very noisy locations.

Despite the effectiveness of the MFSK techniques which this experience proves, it might be claimed that this book will in effect close the stable door after the horse has bolted, since there is a current (and long standing!) prediction that h.f. radio as a means of communication is obsolescent, but the author would counter this argument in two ways.

First, many of the new applications brought about by the changes in the use of h.f. described above are characterised by operation with low transmitter powers and non-optimum aerial systems, in which the role of the demodulation process in establishing the performance of the link is dominant. (This book advances the postulate that there are many applications in which ancillary techniques are less effective than is sometimes claimed in the literature.)

Secondly, although MFSK has been investigated mainly in the field of h.f. telegraphy there is no reason why the same techniques should not be applied in other fields in which digital communication is required by means of signals which suffer distortion in the frequency or time domain or are subject to high levels of noise. For instance, an MFSK system has already been used experimentally for telegraphy by a microwave signal bounced from the surface of the moon, and some tentative investigation has been carried out into applications to underwater acoustic telemetry. With these facts in mind, this book has been written with the primary aim of encouraging the future use of MFSK, rather than simply as a record of past achievements or designs.

A secondary purpose of the book is to review some of the techniques of h.f. telegraphy in the light of more recent research and advances in communication theory. During the last 20 years, the more modern communication fields of microwave, satellite and space communication have enjoyed the full benefit of a spate of theoretical analysis and experimental investigation while the older-established field has not received so much attention, possibly (in more recent years at least), because of the assumption that it is a dying technology. Much theoretical work on telegraphy principles has also suffered from the belief (held for many years in some universities) that the h.f. ionospheric path can be analysed effectively as a non-dispersive channel with added Gaussian noise, and such minor peccadilloes as fast fading and multipath propagation left to be considered empirically by the equipment designer or (more usually) totally ignored, as being too ill-defined to be a suitable subject for mathematical treatment. Such work as has been done on systems to combat multipath has often fallen into the opposite trap, assuming that this is the only problem. If there is one statement that can be made firmly and confidently about the design of an h.f. telegraphy modem, it is that the optimum system must involve a compromise between conflicting requirements.

The author hopes to demonstrate that useful limiting generalisations can be made concerning ionospheric effects and that design around these can be effective, and furthermore that a relatively detailed examination of such phenomena as Rayleigh fading (see Chapter 4) can provide new insight into the causes of transmission errors and hence give some clues as to methods of combating them. Despite the implications of this approach, the aim is not to add to the existing mass of general communication theory, but rather to indicate how simple principles based on such theory may be used to solve very practical communication problems. Nevertheless, it is hoped that the presentation is such as to encourage a more thorough investigation of multi-frequency techniques from the purely theoretical point of view, particularly in those areas where the limitations of the author have necessitated an empirical approach. An attempt has therefore been made to strike a balance between the theoretical and practical. In discussing the concepts of signalling, error coding, and so on, the rigorous approach of the mathematician has sometimes been rejected in favour of simplified and approximate analysis, with physical explanations and generalised conclusions which are easily appreciated and used by practising engineers. A more academic approach may be derived from this and from standard published works by those who wish to study MFSK in the concept of the wider theory of signalling. Some basic knowledge of communication theory and practice (particularly in the field of h.f. telegraphy) is assumed, and lengthy repetition of concepts and analyses discussed in standard textbooks is avoided.

A further objective of the book is to bring to notice some of the circuit techniques involved (such as the matched filter described in Chapter 9 and the synchronisation system in Chapter 8) which may have uses in other fields. Whereas these may contain nothing fundamentally new, they have often been described only in terms of theoretical concepts or narrowly specific applications. In describing circuit designs, reference to specific components or logic families has been largely avoided and the discussion limited to schematics and basic principles, since information specific to detail design is often of little help to anyone wishing to apply that information in another field, and also because with the present day explosion of technology, circuit details rapidly become obsolescent.

The approach to multi-level error coding described in Chapter 11 is rather empirical but simple to apply to practical cases, and it is hoped that the results reported may lead to the investigation in depth that the subject deserves.

The author was introduced to the basic concepts of MFSK in 1957 (by which time a series of empirical experiments had shown that the principles were capable of providing a pronounced improvement in telegraphy accuracy) and since then has been responsible for developing the theoretical basis and superintending the design of successive generations of the Piccolo equipments founded on these principles. The most recent of these (Piccolo Mk 6) is now (1982) entering quantity production and will be marketed commercially, so this would seem a particularly appropriate time to collate the experience of the last 20 years and make it generally available.

1.2 Early MFSK equipment

After the Second World War it was anticipated by the UK FCO that the amount of diplomatic telegraph traffic across the North Atlantic would be likely to rise rapidly in the years to come. Transatlantic cables were at that time few and were suspect, as being too easily disrupted by accident or deliberate sabotage, and h.f. telegraphy was the only available alternative. During the late 1940s a multiplexed FSK system was developed and produced by the Diplomatic Wireless Service (DWS *). In addition, the chief engineer of the day undertook a longer term investigation with a view to the improvement of signalling techniques. A series of empirical experiments culminated in the construction, by 1953, of the Electronic Multiplex Unit (EMU) Mark 2 [1–3]. This was a multi-tone telegraphy system conveying 15 channels of Teletype in time-division multiplex, using 32 audio tones (one for each character of the Teletype alphabet), matched filter reception and orthogonal frequency spacing, in fact all of the fundamental principles of what later became known as multi-frequency shift keying (MFSK). The technology of the day (valves, electromechanical relays etc.) did not permit the required high standard of reliability to be attained and the equipment never went into service but laboratory measurements indicated a pronounced improvement over conventional techniques under poor signal-to-noise conditions.

In 1956 the author (who at that time had had no previous experience in h.f. communication engineering) was asked to head a small team to design a single-channel system based on the principles investigated in EMU, but intended for point-to-point operation from UK to a number of British Embassies. The equipment was to be produced in medium quantities, and it was essential that it should be reliable enough to be viable for continuous operation under tropical

*See Glossary of terms and abbreviations.

conditions with the very limited servicing facilities available at a small embassy. To achieve this aim it was necessary to reject valve techniques in favour of germanium transistors, which at that time were beginning to be available in commercial quantities.

Work began in July 1957 and the first experimental version was tested from Madrid to UK in 1959. The parameters were then modified in the light of that experience and the equipment designed for pilot production. Twelve units were made, and the first *Piccolo Mark 1 [4] (so-called because of the sound of the signal) was installed experimentally in New Delhi in October 1962. The war between India and China was then in progress and the resulting heavy telegraph traffic at the post overcame any administrative problems and forced the use of the experimental equipment on 'live' messages. The superior effectiveness of the new system was proved beyond doubt. It cleared in a few hours a backlog of traffic which had accumulated despite the strenuous efforts of the operators of two hand-speed Morse channels operating as near to 24 hours per day as propagation conditions would permit.

Much was yet to be learned of the practical aspects of the design, and the pilot Mark 1 design was rapidly succeeded by the fully production designed Mark 2, which a year or so later was refined into the Mark 3. Both designs were manufactured in the Departments Production Unit. About 220 units of the Mark 2/3 were made and installed, so that by 1975 virtually no Morse links (other than temporary or emergency links) were operating within the network [5, 6]. At the same time that the Piccolo units were brought into operation, the obsolete a.m./c.w. radio equipment was replaced by state-of-the art units in which all radio frequencies were derived by frequency synthesis from high stability crystal oscillators, allowing single-sideband (SSB) demodulation with no automatic frequency control or manual correction. This method gave a decided improvement in performance over the a.m. system used in the early experiments.

During the early 1970s the author undertook an investigation into the basic principles and theory of MFSK [7] and as a result, when the obsolescence of the Mark 2/3 equipment necessitated a new version, the Piccolo Mk 6 was developed using contemporary technology and different signal parameters as described in Chapter 6. The circuit design and development of this unit took place in 1978–81, circuit principles being developed in the Development Section of the CED and the production design being carried out by Racal Communications Ltd., Bracknell, Berkshire, England. The resulting units began to be installed in service on the FCO network in 1982 and are available commercially as the Racal equipment LA 1117 [8].

Although this mainstream development was being carried out specifically for FCO purposes, many efforts were made to interest other communication authorities, including demonstrations to Armed Services' officers and representatives of manufacturers of communication equipment, and some trials carried out at

---

*See Glossary of terms and abbreviations.

their request [9, 10]. Although initial interest was very slight, the proven success of the Piccolo equipment in FCO service has led to the co-operation of the British Ministry of Defence in the specification of the Piccolo Mk 6 in order to ensure compatibility with existing and future Services equipment and tests are in progress with a view to its use on strategic links.

There were also a number of projects in which the Department co-operated with other organisations in the design of MFSK equipments for specialised applications. The first of these (shortly after the first Piccolo Mk 1 went into service) was the design by the Signals Research and Development Establishment (of the British Army) of a microwave telegraphy link reflecting from the surface of the moon. First experiments using conventional FSK techniques at 800 baud had failed, owing to the long echo on the signal (about 12 ms) and violent selective fading. All available information on the Piccolo design was given, and eventually an experimental MFSK system was constructed and operated successfully [11] (see *Section 12.6*). The project was an encouraging example of the power of MFSK techniques and their adaptability to unusual communication problems.

The second such project led to co-operation between DWS, *The Daily Telegraph* newspaper, and Cunard Ltd. to produce the first true 'newspaper by radio' for the Cunard luxury liner Queen Elizabeth 2. The paper was fully edited and assembled in London and the complete text and compositing data sent nightly over the existing marine telegraphy network, to be automatically composited and printed by computer-controlled photosetting equipment on the ship. Again the early experiments using FSK techniques with error coding had failed to give the required standard of accuracy, and following convincing proof of the superiority of MFSK techniques (the first experimental two-way trip across the Atlantic reported 2 errors in 300 000 characters – without error coding), a Piccolo receiving system was loaned to Cunard Ltd. (through a licensing arrangement with Marconi Company Ltd.) and installed on the QE2 on her proving voyage in April 1969, the Piccolo signals being conveyed at audio frequencies via the Post Office telephone system from Fleet Street in London to a special synthesised SSB transmitting facility at the BPO Marine h.f. communications station at Portishead, near Bristol, England. The newspaper was a complete success and continued to be a feature of life on the ship until rising costs forced the withdrawal of the facility after about 10 years of successful operation [12].

The next project originated from an approach by the National Institute of Oceanography for assistance in the design of a telemetry system to operate between a large (40 tonnes) unmanned data-collecting buoy (D.B. 1) anchored about 300 km offshore, and the mainland. The data were to be transmitted at fixed times, on a single fixed frequency over a low-power h.f. link and without a return channel. CED (as it had by then become) co-operated with Racal Electronics Ltd. in the design of a special MFSK system (see Section 12.3), which was completely successful 'setting new standards in marine telemetry' (to quote the users report) [13, 14]. Racal purchased limited design rights and later produced and installed a small number of telegraphy equipments based on the same design (under the Racal equipment number LA 2028).

As part of the involvement in the Data Buoy project, a report was submitted [15] to the Commission of European Communities recommending MFSK techniques for marine h.f. telemetry application, and this led to co-operation with the Christien Michelsen Institute in Bergen, Norway, on comparative trials between an MFSK and a binary system as discussed in Chapter 7.

1.3 MFSK and the communications industry

The early history of telegraphy is dominated by the use of the Morse code, the sending 'mechanism' consisting of a single contact operated by hand and the 'demodulation system' a simple transducer giving visual or aural indication of the passage of a current. Both are predominantly binary processes, and the later extension to sending contacts operated by machinery, with electromagnetic decoding and printing devices, and from d.c. signalling to audio and radio links, did not bring about any serious challenge to binary principles. The reliability and simplicity of binary devices was, of course, a major factor in maintaining this situation, and it is also important to note that communication theory (in the sense of the analysis of the process of communication itself, as distinct from the technology by which communication is achieved) was not seriously applied to radio links until the early 1950s.

The reasons for the failure of the communications industry to accept and develop the principles of Piccolo in the early 1960s are many and complex. There are no doubt wide variations of viewpoint, and the following analysis represents the personal opinions of the author.

To begin with there were some new and very real technical problems. Amplitude and frequency modulation techniques had been rejected at an early stage, and, because of the narrow bandwidths of the tone filters, single-sideband modulation required radio frequency stabilities from 20 to 50 times better than current good commercial practice. This could only be met by using radio frequencies in the transmitting and receiving equipment generated entirely by synthesis from stable standards. Such techniques were still in their infancy and were mostly confined to expensive laboratory test gear. The insistence by the Piccolo designers that they should be extended to communication equipment for use in the field anticipated the natural evolution which has now enabled the requirement to be satisfied by most good quality commercial radio equipment. Another valid objection was the complexity of the demodulating circuits. The experience with the EMU demonstrated that the electronic techniques of the 1950s (based on the valve and electromechanical relay) were simply incapable of providing the required degree of reliability and repeatability when used in quantity and operated under field conditions, and the resulting items of equipment were bulky, heavy and uneconomic. It required the development of the transistor to make equipment containing circuits equivalent to several hundred valves into a practical proposition, but the attitude at that time was to use transistor techniques as a replacement for the valve in circuits based on existing principles, rather than to see them as a means to progressing to more complex concepts.

There were also more subtle factors, attributable to the commercial structure of the telegraphy industry of the day. The world-wide intercontinental trunk telegraphy networks operating ITU-recommended systems dominated the development of h.f. communications systems generally, and the current demand in that field was for the ultimate in spectrum conservancy to maximise the traffic capacity of existing installations. The authorities were not interested in a totally new technique which promised no improvement in that direction, and preferred to look to Error Detection Coding to improve traffic quality. As one colleague expressed it at the time, it was like trying to sell Land Rovers to a company operating a fleet of motor coaches – and the analogy is remarkably appropriate. It was also believed, quite understandably, that it would be impossible to obtain international agreement on such an unorthodox system within a reasonable time scale.

The second major application of h.f. telegraphy was the marine networks to naval and merchant shipping. These were virtually dedicated to hand-sent Morse operation as providing the highest possible circuit availability at the lowest cost with the simplest equipment. There seemed to be a current belief that it was not possible to design an automatic telegraphy system which could improve on the human brain as a means of interpreting weak signals in noise or when badly distorted, and it proved difficult to convince some authorities that it was possible to signal accurately at 100 words per minute under signal conditions where expert Morse operators were unable to communicate at a tenth of that speed. For most ships the amount of traffic was normally well within the capacity of hand Morse (or was constrained to be so), and cost-conscious shipping lines did not foresee the considerable rise in traffic demand to even small terminals when suitable facilities are available. (For instance, it is not uncommon nowadays for all routine administrative transactions of a freight ship to be exchanged while it is still at sea, thus reducing time in port.) Without the support of either the intercontinental trunk networks or the marine telegraphy networks, commercial interest was understandably absent.

This polarisation of the world of h.f. telegraphy, dominated on the one hand by high-power heavily-multiplexed point-to-point links operating machine telegraphy with high traffic density, and on the other by low-usage hand Morse operation to very small and simple stations, obscured the fact that many users really require a compromise between the two. Compact, reliable telegraphy terminals operated by non-specialist non-technical staff and giving good-quality single-channel teleprinter-speed communication have a very wide range of applications, as is proved by the success of the Telex network. Third world and emergency networks over difficult terrain, back-up links for cable and satellite networks, and scientific and commercial telemetry from unmanned stations, are all situations where the simplicity and relatively low capital cost of h.f. telegraphy may outweigh the problems of a variable and difficult communication channel. The use of new techniques on such networks does not require international acceptance (except to satisfy requirements to minimise interference – and in this respect MFSK is an improvement over many other systems) and the publication of a CCIR Report on MFSK [16] suggests that the principles have received some measure of recognition.

As the supply of well-trained and experienced radio operators generated by the Second World War has inevitably disappeared. the hard economics of the present day has forced the recognition of the high manpower cost of such an individual and personal skill. When Morse operation is replaced by machine telegraphy the weaknesses of conventional FSK techniques over the difficult medium of radio become only too evident. This brings us to the final major objection raised to the development of Piccolo, the conviction current at the time that satellite communications would rapidly replace all other forms of long-range signalling and that h.f. was therefore obsolescent and did not justify further investigation. In fact while the 1960s saw considerable movement towards satellite communication, this was mostly in the large-capacity trunk networks, and even in the early 1980s the use of satellites to provide relatively low cost, medium-capacity links to small terminals is still far from extensive. Experience with the new medium inevitably brought about a more rational and balanced assessment of the relative advantages and drawbacks of the two techniques so that the 1970s saw a considerable swing back to h.f. communication and there is a general belief that the medium will continue to be used extensively for the foreseeable future.

By the time that the force of these counter-arguments was becoming more evident, and the experience of the FCO and independent trials had established beyond doubt that an MFSK system could be technically viable, the situation was that the only Piccolo units in existence in quantity were those made by the FCO for their own use. These were already obsolescent in circuit techniques, rather specialised in facilities, and not suitable for commercial manufacture, so that despite increasing interest shown by organisations with difficult h.f. problems, no further extension of Piccolo systems could take place until FCO requirements justified a complete redesign and manufacture in commercially viable quantities.

The Mark 6 Piccolo has been deliberately designed as a flexible, general purpose telegraphy modem. This book attempts not only to present the background knowledge and experience embodied in that design, but also to show how that knowledge may be applied to other communication problems, and indicate directions in which it may usefully be extended by further work. The future alone can show whether or not MFSK techniques will be accepted as an additional weapon in the fight for better communications.

1.4 References

1 ROBIN, H. K., and MURRAY, T. L.: 'Receiver systems'. UK Patent 746 741, 14 Dec 1949
2 ROBIN, H. K.: 'Improvements in and relating to electric pulse signalling systems'. UK Patent 746 743, 30 Dec 1949 and 11 Dec 1950
3 ROBIN, H. K., and MURRAY, T. L.: 'Improvements in and relating to receiving apparatus for frequency shift telegraphy'. UK Patent 854 201, 18 June 1957
4 ROBIN, H. K., BAYLEY, D., MURRAY, T. L. and RALPHS, J. D.: 'Multi-tone signalling system employing quenched resonators', *Proc. IEE,* 1963, **110**, (9), pp. 1554–1568
5 BAYLEY, D. and RALPHS, J. D.: 'The Piccolo 32-tone telegraph system', *Point to Point Telecommunications*, 1969, **13**, No. 2, pp. 78–90
6 BAYLEY, D. and RALPHS, J. D.: 'Piccolo 32-tone telegraph system in diplomatic communication', *Proc. IEE*, 1972, **119**, (9), pp. 1229–1236
7 RALPHS, J. D.: 'The application of MFSK techniques to h.f. telegraphy', *The Radio & Electronic Engineer*, 1977, **47**, No. 10, pp. 435–444
8 RALPHS, J. D.: 'An improved Piccolo MFSK modem for h.f. telegraphy', *The Radio & Electronic Engineer*, 1982, **52**, No. 7, pp. 321–330.
9 SCHEMEL, R. E.: 'An assessment of Piccolo, a 32-tone telegraph system', SHAPE Technical Centre Memorandum STC TM-337, 1972, File Ref. 9980, DRIC No. P186242
'Assessment trial of Piccolo telegraph modem', Army School of Signals Assessment Trial Report 248/78, Feb 1978
10 WESTON, M. A.: 'Microwave moon communication at high digit rates', *Proc. IEE*, 1968, **115**, (5), pp. 642–651
11 RAWCLIFFE, A.: 'A newspaper for QE2', *Point-to-Point Communication*, May 1971
12 RUSBY, J. S. M., KELLEY, R. F., WALL, J., HUNTER, C. A. and BUTCHER, J.: 'The construction and off-shore testing of the UK Data Buoy (DB1 Project)', Proceedings of the Technical Session (Instrumentation and Communication), Oceanography International 1978
13 RUSBY, J. S. M., and WAITES, S. P.: 'The deployment and operational performance of the DB1 Data Buoy', Ocean 80 IEE, forum on Ocean Engineering 1980
14 RALPHS, J. D.: 'Application of MFSK techniques to the COST 43 Project'. Submission by UK to the Data Transmission Sub-Group. COST 43 Project. Commission of European Communities, Brussels 1974
15 CCIR: 'Multi-frequency-shift keying techniques for h.f. telegraphy', Kyoto, 1978, **3**, Rep. 702, pp. 171 ff

Chapter 2

Basic Principles

Note: Throughout this book the term 'multi-frequency signalling system' (or similar) refers to one in which data is conveyed by selection from more than two frequencies. The term does *not* include binary systems (amplitude, phase or frequency modulated) which convey several channels in parallel by frequency-division multiplexing.

2.1 Preliminary considerations

The essential process inherent in digital communication is that at the sending end the sending mechanism makes one or more choices from a number of available alternative signals (referred to in general discussion as 'levels'), that choice depending on the information to be transmitted. The receiving system has an *a priori* knowledge of each of the signals which may be selected, and the received signal, (consisting of the transmitted signal attenuated and distorted by the communication path and with added noise and interference of various types) is inspected in order to determine which of the various alternatives is the one most likely to have been transmitted.

The minimum possible number of levels is two, and the fundamental simplicity of binary signalling has led to its overwhelming predominance. However, there is no justification for the all too frequent assumption that this is the only system, or necessarily the best. The work of Shannon [1] established that, given a channel of fixed bandwidth and signal-to-noise ratio, there is a defined limit to the information rate achievable with negligible errors. Slepian [2] and Viterbi [3] investigated the performance of systems with differing numbers of levels and showed that in general it improved rapidly as the number increased above two, but that the improvement was negligible for large numbers (above about 100). These conclusions can be confirmed in general terms by deductions from fundamental concepts as follows.

The Shannon theorem [1] states:

'By sufficiently involved encoding systems we can transmit information with arbitrarily small frequency of errors at the rate:

$$H = B \log_2 (1 + S/N^2) \text{ bits*/s} \tag{2.1}$$

where H is the maximum information capacity, B is the bandwidth of the signalling channel, S is the mean signal power, and N^2 is the mean thermal noise power in the bandwidth B.

It is not possible to transmit at a higher rate by any encoding system without a definite positive frequency of errors.'

This equation may be recast in more convenient terms (with certain simplifying assumptions) as:

$$B_0 \log_2 (1 + R_0/B_0) \geqslant 1 \tag{2.2}$$

where

$$B_0 = B/H$$
= normalised bandwidth in Hz/bit/s

$$R_0 = \frac{S}{H} \frac{B}{N^2}$$
= normalised signal-to-noise ratio

$$= \frac{\text{signal energy per bit}}{\text{noise power per Hz bandwidth}}$$

Shannon's phrase 'sufficiently involved encoding' implies that to approach this limit involves increasing the decoding delay (in the general sense of the amount of information which must be accumulated by the receiver before any of it can be interpreted), and other work confirms this concept. Now the larger the number of alternative signals available to the sender, the greater the amount of information in each choice (see the definition of 'bit' in the 'Glossary of terms'), and since for a constant data rate the duration of each signal is increased, an increase in the number of levels implicitly involves an increase in decoding delay and a consequent improvement in performance. However, the relationship between the information per level and the number of levels is logarithmic, so that the rate of improvement is less for a large number of levels.

Accepting the theoretical advantages of a multi-level system, and recognising the very practical requirement for radio telegraphy that the signal should be

---

*See glossary of terms for definition of the 'bit'.

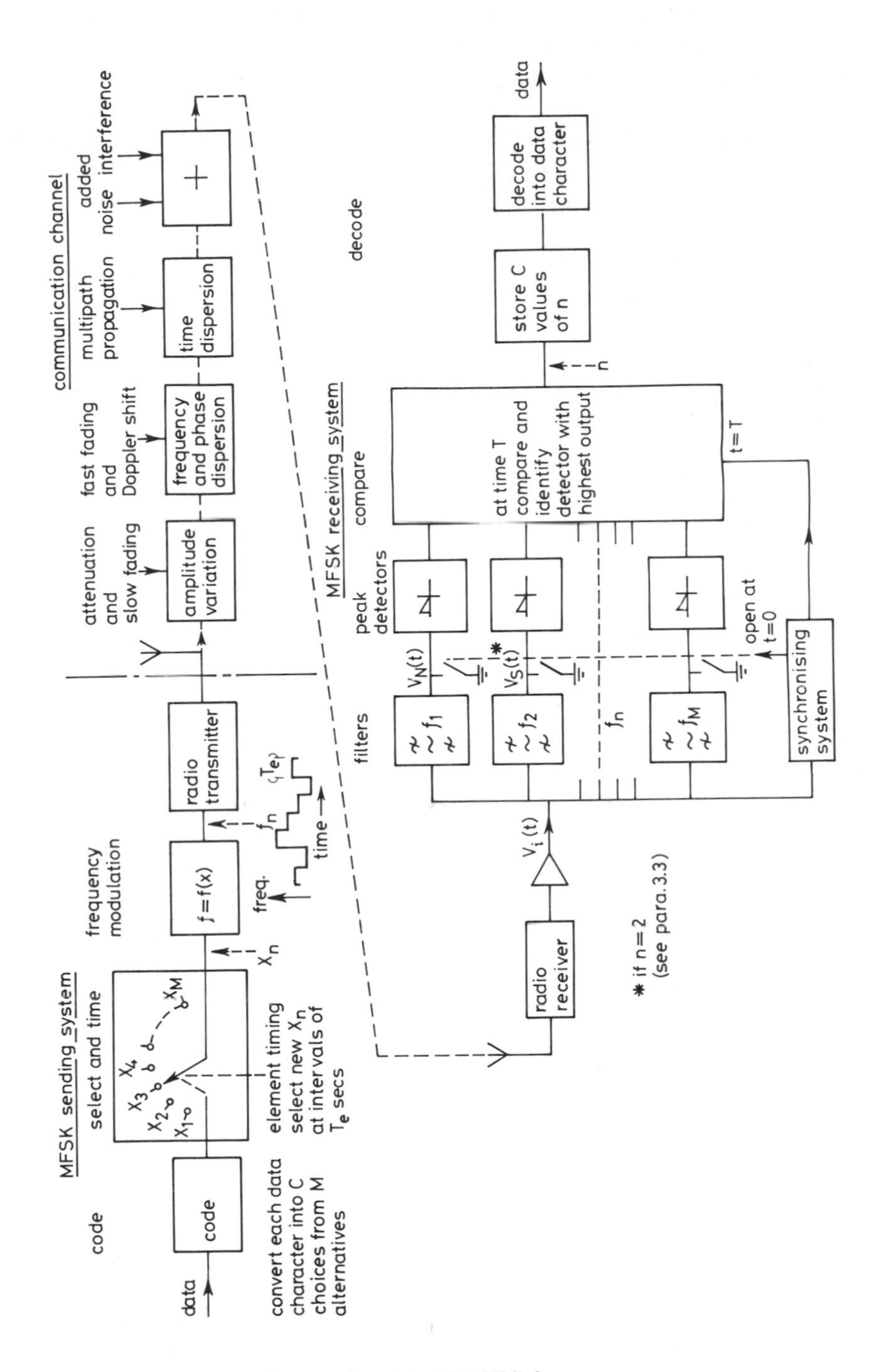

Fig. 2.1 *Block schematic of assumed model of MFSK link*

restricted to a relatively narrow band and therefore the information must be carried as a modulation on a sinusoidal waveform, one is faced with a choice between amplitude, frequency or phase modulation [4]. In the case of binary signalling, the superiority of frequency modulation (frequency shift keying) over amplitude modulation (on–off keying) has been established for many years both in theory and in practice. A study of phase shift keying shows that as the number of possible signal levels increases, the differential phase angle between any two of them decreases, and therefore the signal-to-noise ratio required in order to make a reliable assessment of the phase must be increased. For a communication system which must transmit the maximum amount of information in a restricted bandwidth, but where a high signal-to-noise ratio is available, a multi-level phase modulation system may be worth consideration, but where noise is a governing restriction on the accuracy of communication, a frequency modulation system would seem to be preferable.

Assuming, then, that the sending process consists of selection from a number of alternative signal frequencies, these may be transmitted as a single frequency at a time (multi-frequency shift keying), or as two or more signals transmitted simultaneously (frequency permutation keying) as investigated by Schneider [5]. The latter system may well be indicated for use over a stable low-noise channel conveying data with a large alphabet (see Section 2.5) but it is doubtful whether it could be usefully applied in most practical cases for a number of reasons. For instance, if two tones are transmitted simultaneously the element duration is doubled (for a constant data rate) and, from Section 3.2, this will give a 3 dB improvement in effective signal-to-noise ratio, but in Section 7.7 it is shown that if two tones simultaneously modulate an SSB transmitter there is a reduction of 6 dB on the radiated power of each tone. The simultaneous process therefore represents an effective loss of 3 dB in signal-to-noise ratio compared with single sequential tones. Furthermore, in a permutation system the two tones transmitted simultaneously cannot be of the same frequency, so that for a fixed number of alternatives, the number of available frequencies must be increased, increasing complexity, cost and bandwidth. Permutation modulation will not be considered further, and discussion will be limited to MFSK.

2.2 Multi-frequency shift keying

In an M-level multi-frequency system the available alternative signals consist of M different tone frequencies, one of which is selected at a time for a fixed duration of time, referred to as a signalling element. The amplitude and time duration of each element is kept the same, since this means that the energy per element is constant (which is one of the criteria for an optimum system), and while it is not theoretically essential, practical considerations suggest that the transition between one frequency and the next should be made without any sudden jump in phase. In effect, therefore, the transmitted signal is a single continuous sinusoid which changes frequency at fixed intervals of time.

Such a signal may be modulated onto a radio carrier by any of the normal methods, and if limited to a 3 kHz bandwidth may be treated in the same manner as a speech waveform, such as transmission by independent sideband a.m. (with or without carrier reduction or suppression), or modulation onto sub-carriers. The simplest concept, however (and the most efficient modulation for a single signal), is if the signal as described is transposed directly to the required radio frequency, giving effectively frequency modulation of a carrier by a stepped waveform as shown in Fig. 2.1. This may be designated by the CCIR classification F1B FN [6]. The most effective method of demodulation is by conventional SSB techniques, mixing with fixed-frequency oscillators to transpose the signal back to baseband frequencies. This is the technique assumed throughout this book unless otherwise stated.

Probably the first proposal for, and mathematical analysis of, a multi-frequency tone signalling system of this type was presented by the Russian communication engineer V. A. Kotel'nikov in his doctoral dissertation at the Molotov Energy Institute in Moscow in 1947 [7], and despite the many advances in communication theory since then, this analysis is still exceptionally clear, primarily because he considered limited practical cases rather than attempting to establish more general relationships. When this work became available to the author in about 1961 it was reassuring to realise that it gave complete support to the very practical and empirical approach of the EMU experiments since the system proposed by Kotel'nikov was virtually identical in its fundamental principles with that developed in the EMU and perpetuated in the Piccolo Marks 1, 2 and 3.

Multi-frequency systems have been used for various line signalling and control applications (as, for instance, the Coquelet system), but the requirements are so different from those considered here that there is little point in discussing them. Applications of multi-frequency signalling (other than MFSK) to radio telegraphy have largely been limited to one or two experimental systems [8].

For analysis purposes an MFSK link may be expressed diagramatically as shown in Fig. 2.1. A signal source generates a sine wave at a frequency controlled by a function x which may be selected to be any one of M values, each of which will generate one of a series of frequencies $f_1, f_2, f_3 \ldots f_M$. Each selection of x is maintained for a period of T_e seconds and the change between frequencies is assumed instantaneous and transient free. It can be shown that an optimum detection system for such a signal involves two fundamental practical techniques, those of 'matched filter reception' and 'orthogonal signals' which are discussed under those headings in many modern books on communication theory [9] and have been used in a number of modern signalling systems such as Kineplex, ANDEFT and Kathryn (see Chapter 7).

2.3 Matched filter reception

A suitable reception system for such a signal would consist of a number of bandpass filters, one for each possible tone frequency, on the assumption that the filter which gives the greatest response to the signal is the one corresponding to the frequency transmitted. The question then arises, what is the best type or design of filter to give the clearest indication of the correct signal, i.e. with the least probability of error? The answer is that each filter should be 'matched' to one of the possible signals in the sense that it will produce the maximum ratio between the response to that signal and the response due to noise or other signals. Analysis of matched filters shows that the maximum immunity to white Gaussian noise is obtained if the filter design is such that a sharp impulse input produces a transient response identical to the shape of the signal waveform itself. (An alternative approach uses 'cross-correlation', in which the signal is compared with stored replicas, but this leads to the same overall performance and is not pursued here).

Consider now a resonant *LCR* circuit tuned to a frequency f, and apply to it a very narrow pulse of current as shown in Fig. 2.2. The circuit will be shock excited to an initial response amplitude depending on the amount of energy in the pulse, followed by an oscillatory 'ringing' of the tuned circuit at the frequency f, dying away as the energy is dissipated in the resistance. If the resistance is increased, the oscillation dies away more slowly until, if the circuit is lossless, the oscillation continues indefinitely at the initial amplitude. It follows that one

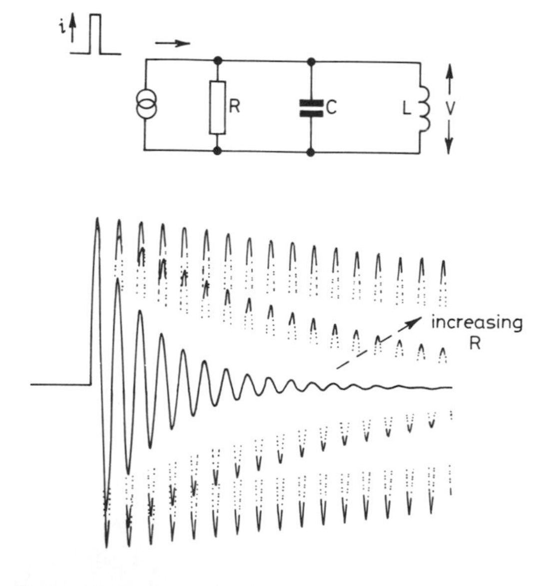

Fig. 2.2 *Impulse response of a resonant circuit*

approach to the realisation of a matched filter for a sine wave of constant amplitude is by the construction of a resonant circuit with no losses. Practical approaches to this, and other methods of achieving the same end are discussed in Chapter 9.

If energy from the preceding signal element is not to contribute to the output of the filter it is necessary for the outputs of all filters to have zero energy at the beginning of each new element. Consider then the use of a circuit such as shown in Fig. 2.3 as a matched filter. With a finite value of R, let the signal be applied at the resonant frequency of the circuit. When the switch is opened, the envelope of the filter output will build up exponentially to a limiting value. If the signal is off-tune from the filter frequency and the process is repeated, the initial rate of build-up is the same but there is a transient oscillation of the envelope before the output

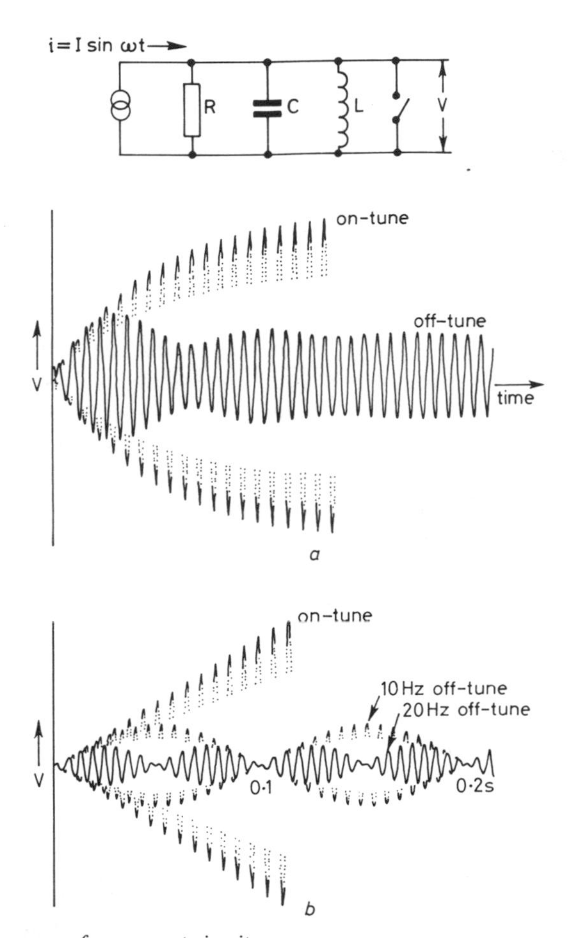

Fig. 2.3 *Tone response of a resonant circuit*
a Response with finite losses
b Response with zero losses (R = infinity)

settles down to a lower final level according the selectivity of the circuit as shown in Fig. 2.3*a*.

If the losses are eliminated (R = infinity), for an on-tune signal the response will rise linearly from the time that the switch is opened and will continue increasing indefinitely (in effect the circuit integrates the signal). For an off-tune frequency the sinusoidal 'transient' response continues indefinitely, with a modulation depth equivalent to 200% amplitude modulation, so that the envelope passes through successive zeros with a phase reversal of the signal at each zero. These effects are proved in Section 3.1. The important characteristic is that the time interval between successive zeros is equal to the inverse of the frequency difference between filter and the signal. In the examples shown in Fig. 2.3*b*, for a 10 Hz frequency difference the zeros are separated by 100 ms, and for a 20 Hz difference by 50 ms. Since the response starts from zero at the instant the switches are opened this is the time of the first zero in all cases.

2.4 Orthogonal spacing

This important characteristic of a matched filter for a burst of tone is used in MFSK systems to reduce the bandwidth occupied by the system. From the basic concept it is evident that the incoming signal should have the maximum effect on the filter which is tuned to it, and the minimum (preferably zero) effect on the output of all the other filters. If this were attempted by conventional filter design, each tone frequency would need to be separated from the frequencies above and below it by an amount sufficient to allow the selectivity of the filter to reject these frequencies, and the tone duration would need to be sufficient to allow all the transient responses to die away and leave the steady states only. Attempts to reduce the bandwidth of the system by using very high selectivity filters would then slow down the signalling speed by needing a long settling time. By using the 'transient' characteristics of the matched filter, both these problems are solved neatly and effectively as follows.

Let the filters in Fig. 2.1 be of the matched filter design described above, and separated at intervals of 10 Hz, and consider the response to a 500 Hz tone of the filters tuned to 490, 500, 510 and 520 Hz. As we have seen previously, the 500 Hz filter will build up linearly from the instant that the switches open. The 490 Hz filter response will have the same initial slope and an oscillatory response which goes through zero at a time 100 ms from that instant. The 510 Hz filter response will have an envelope identical to that of the 490 Hz. The 520 Hz filter response will be lower in amplitude and will go through zero every 50 ms, so that it too will be going through zero 100 ms from the origin. In short, it can be seen that *any filter centred on a frequency which is separated by a multiple of 10 Hz from that of the applied tone will be going through zero 100 ms from the opening of the switches.* To detect the frequency of the applied tone with maximum confidence it is only necessary to simultaneously compare the envelopes of all the filter outputs at an instant 100 ms after the switches open, when the 'wanted' filter will show an output, while all other filters will be at zero.

This is the principle of 'orthogonality', which may be expressed more formally as follows. Let information be transmitted by a single selection of one signal from a series of M alternatives:

$$X_1, X_2, X_3, \ldots X_M$$

In the receiving system let this be applied to a series of M detectors so as to produce the responses:

$$Y_1, Y_2, Y_3, \ldots Y_M$$

Then for an orthogonal system, a signal X_i will produce the responses:

$$Y_i = 1 \qquad Y_n = 0 \text{ if } n \neq i$$

(The term bi-orthogonal is occasionally met. A system is bi-orthogonal if $Y_i = 1$ and $Y_n = -1$ for $n \neq i$).

All of the MFSK systems discussed in this book are nominally orthogonal, as are some, but not all, of the binary systems referred to in Chapter 7.

A complete cycle for the detection of an element can then be studied with reference to Fig. 2.4. At a time known to be after the beginning of the signal element, all the switches are simultaneously opened and the signal sampled in each filter for a period T. At the end of the period the envelopes of the filter outputs are compared, and a decision made as to the highest, this decision constituting the output of the system. Immediately the decision is made, all switches are simultaneously closed and the energy in the filters is discharged to zero during time T_d, referred to as the 'dead time'. The cycle is now completed and the system is ready to operate on the next signal element. This type of

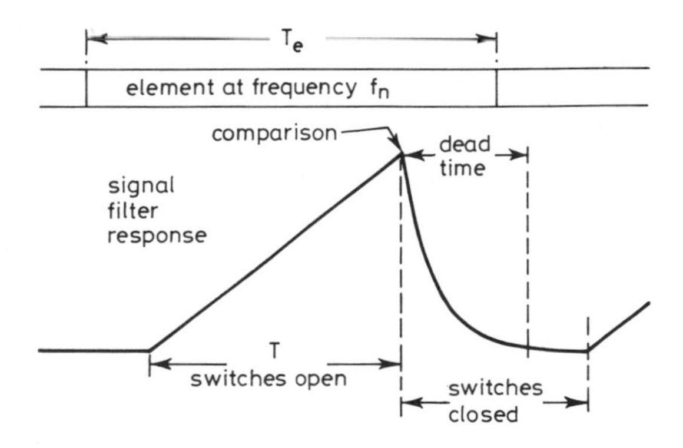

Fig. 2.4 *Timing of receiving process*

matched filter is frequently referred to in the literature as an 'integrate-and-dump' filter for obvious reasons. The time for which the switches are closed is referred to as the 'guard time'.

Thus the matched filter, which in itself gives the maximum probability of detecting correctly a single tone in white noise, may also be given the maximum protection against signal frequencies other than the one to which it is tuned, if the frequency difference between any two possible signal tones is made an integral multiple of the inverse of the sample time, i.e. if the signal frequencies are spaced at intervals of n/T Hz where n is an integer (usually $n=1$). The complete action may be seen by reference to Fig. 2.5, which shows the envelopes of the responses of filters at f_1 and f_3 to a signal sequence f_1, f_2, f_3, f_6.

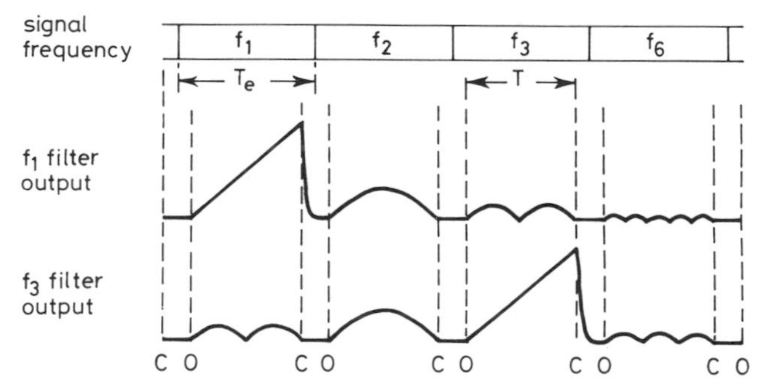

Fig. 2.5 *Responses of two matched filters to a tone sequence*
O Switch contacts open C Instant of comparison

2.5 Choice of parameters

One point which should be emphasised from the beginning is that in most cases the design of an MFSK system is closely bound up with the characteristics of the actual data source itself. This is in contrast with the situation in binary signalling, where almost the only characteristic of a communication system which may limit the type of data to be sent over it is the maximum modulation rate it can handle. A binary link designed for a modulation rate of say 1200 Bd can, in principle, transmit and receive binary data in any format within this limitation, whether the data are synchronous or asynchronous, continuous bit stream or discrete bytes, in different codes, or at a lower information rate (although in the last case it may not give optimum performance). On the other hand an MFSK modem is normally designed to handle data in bytes of specified size, and at a specified rate. This tends to make an MFSK communication channel less flexible than a corresponding binary system, although measures may be taken to increase this flexibility.

In designing an MFSK system therefore, the first step is to analyse the data source itself. There are various possible approaches but probably the simplest is to start from the size of alphabet required, The 'alphabet' is defined for this purpose as the number of different symbols (or 'characters') in the given data input. For instance, the ITA-2 telegraphy code requires an alphabet of 32 characters, the ITA-5 (ASCII) code requires 128, a frequent requirement is a number code conveying the 10 decimal digits, while a telemetry link required to pass numerical data in the form of 3-figure measurements (000 to 999) could be expressed in decimal (requiring an alphabet of 10), or in an alphabet of 1000. Note that in some cases synchronising or error coding requirements may necessitate an increase in the size of alphabet but these are ignored in this first analysis.

Having settled tentatively on a size of input alphabet A_0, data consisting of a sequence of such symbols is to be conveyed by a sequence of tone elements. A limitation which is most convenient, particularly with large alphabets, is that each symbol should be represented by an integral number of elements, i.e. no element should contain information from more than one symbol. This arrangement also reduces to a minimum the effect of transmission errors, in that one element error can only corrupt one symbol. So let each symbol be conveyed by C elements, each selected freely from M frequencies. The total number of different symbols which can be generated (noting that the same frequency may be sent on two successive elements if required), is:

$$A_e = M^c \tag{2.3}$$

In principle C can have any integral value from 1 (when $M = A_e = A_0$) to $C \geqslant \log_2 A_0$ (when $M = 2$).

The other major parameter is the element duration. The requirements will presumably specify a required symbol rate, the inverse of which gives the symbol period T_0, and since C elements must be transmitted in this period then:

$$T_e = T_0 / C \tag{2.4}$$

The above factors can be combined to give a required data rate:

$$H_0 = \frac{\log_2 A_0}{T_0} \text{ bits/s} \tag{2.5}$$

and a system data rate:

$$H_e = \frac{\log_2 M}{T_e} \text{ bits/s} \tag{2.6}$$

It is evident that $H_e \geqslant H_0$ and if this is not an equality the code includes 'redundancy' and there is a consequent waste of signal energy given by:

$$E = \left(\frac{H_e}{H_0} - 1\right) 100\% \tag{2.7}$$

It is informative to use as an example a system requiring a large alphabet, so let us assume that there is a requirement for a telemetry system in which each input byte of data is a 3-digit decimal number (from 000 to 999 inclusive) and 5 such bytes are required to be transmitted each second. Then:

$$A_0 = 1000 \text{ symbols (or bytes)}$$
$$T_0 = 0.2 \text{ seconds}$$
$$H_0 = \frac{\log_2 1000}{0.2} \simeq 50 \text{ bits/s}$$

A table can then be constructed, as shown in Table 2.1, of possible systems fulfilling the requirements. The bandwidth is calculated according to Section 3.4.

Table 2.1 *Possible MFSK systems for an alphabet of 1000 bytes at 5 bytes/s*

C	M	A_e	E(%)	T_e(ms)	B(Hz)
1	1000	1000	0	200	5020
2	32	1024	0.34	100	360
3	10	1000	0	66.7	210
4	6	1296	3.75	50	200
5	4	1024	0.34	40	200
6	4	4096	20.41	33.3	240
7	3	2187	11.33	28.6	245
8	3	6561	27.23	25	280
9	3	19683	43.14	22.2	315
10	2	1924	0.34	20	300

From the table it can be seen immediately that some of the systems (such as $C=6$, 8 or 9) involve very high redundancy and may be eliminated on these grounds alone, while high values of M require an extremely wide bandwidth, so that for analysis purposes the choice may be restricted to the four systems ($C=2$, 3, 4 or 10), which are illustrated in Fig. 2.6 to the same time and frequency scales. With a smaller alphabet, the possibilities may reduce to a single system or a choice between two.

The parameters suggested by the alternative systems must then be examined in the light of the characteristics of the communication path over which the signal is to be sent. The primary parameter is the element length. If the element is long, then a high standard of frequency stability will be necessary, and frequency or phase instability in the signal path or equipment may cause a corresponding degradation of signalling accuracy. On the other hand, a system using a short element will tend to be disturbed by variations in the transit time of the signal,or by 'multipath' effects in which two signals are received with different delays, giving an 'echo'. The instabilities of the equipment and the signal path in the frequency and time domains thus determine the upper and lower limits, respectively, of the permissible element length as discussed in Chapter 3. If two or more alternative systems are acceptable, then choice of the optimum can be carried out at discussed in Chapter 6.

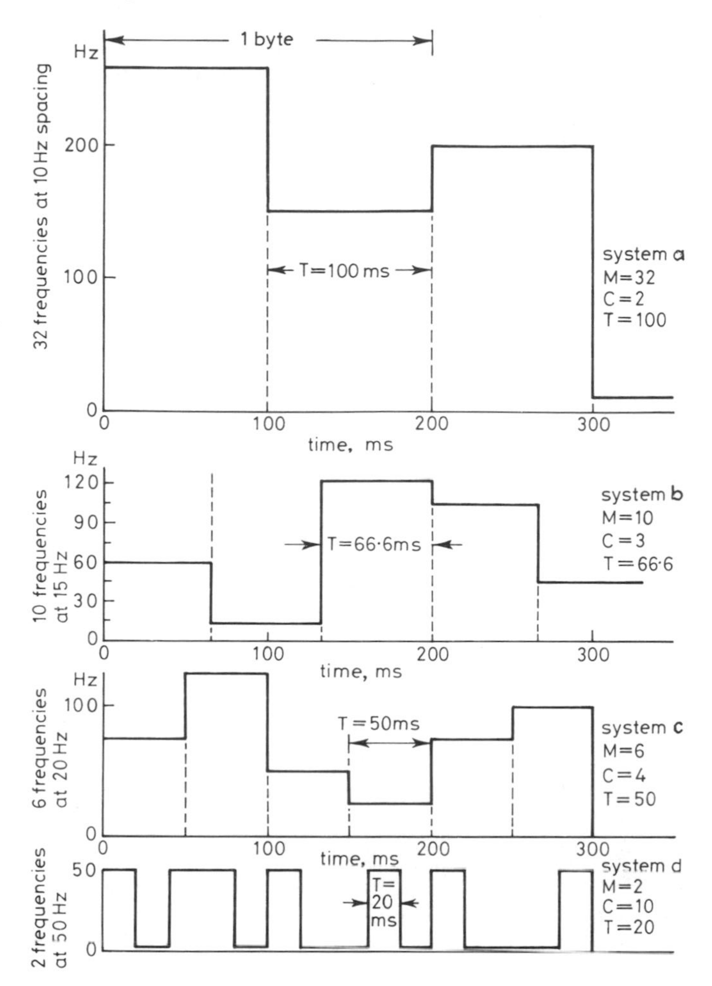

Fig. 2.6 *Possible alternative MFSK systems*
10 bits per byte; 5 bytes per sec

If no suitable system is indicated by this procedure, then it may be necessary to sacrifice the advantages of operating with each character transmitted in integral elements and to reduce the alphabet (in the example above, the data could be sent in two-digit bytes, giving an alphabet of $A = 100$). Alternatively, the alphabet may be increased by signalling in 'bigrams' consisting of bytes containing two successive characters in the input data. The ITA-2 code sent in this way would have an alphabet of $32^2 = 1024$ bytes. If these methods still yield no suitable system, then the original requirements may need to be revised, as being beyond the capabilities of the signalling channel.

2.6 Coherent and Non-coherent Channels and Demodulation

A communication channel may be defined as being coherent if, once the absolute phase of the audio signal has been determined at the receiving end over a limited period, the absolute phase of all possible subsequent signals is known in advance. A coherent reception system is one which takes advantage of this characteristic of the communication channel and assumes a knowledge of the audio phase. A simple method of demonstrating the advantage of using a coherent detection system (where this is possible) is by reference to the description of the quadrature-modulation matched filter in Section 9.2 and Fig. 9.2*a*. If the phase of the signal is known, one of the reference waveforms can be arranged to be in the correct phase, and the other (quadrature) path omitted. The single integrator will then produce a D.C. output proportional to the signal amplitude, with a fluctuation on that output from the noise component. This would constitute a matched filter for a coherent signal. If the phase of the signal is unknown, the quadrature path is required in parallel, and if the signal happens to be in the correct phase for the first modulator the second will contribute no signal output but only a noise fluctuation equal in power to that from the first. Unfortunately many communication channels (and in particular all h.f. propagation paths) are unstable and produce time-varying phase fluctuations, rendering coherent detection impracticable.

Some, but not all, of the advantages of coherent operation may be recovered by quasi-coherent methods, as for example differential phase shift keying, which is discussed in Section 7.5. With the growth of highly complex electronic circuits, it may be possible to apply such techniques to MFSK, using the received phase of one element to establish a reference for the phase of the detection process for the next. However, care should be taken in advocating such principles as discussed in Section 7.4.

Note that even so-called non-coherent detection normally assumes the signal to be coherent over a limited period of time (usually one element), and when this assumption is not met, 'fast fading errors' may occur, as described in Section 4.8.

2.7 Synchronous Detection and Synchronising

A data stream may be defined as being synchronous if, once the time of arrival of one element has been determined accurately at the receiving end, the exact time of arrival of all elements in subsequent characters can be predicted. The conventional mechanical teleprinter keyboard generates an asynchronous signal (the transmitting mechanism being actuated when a key is depressed). A continuous tape reader is quasi-synchronous in that the output characters are sent at a data rate determined by the rate of rotation of a mechanism, but in many cases this is relatively poorly controlled (typically a tolerance of $\pm 1\%$). Reception of such signals is normally carried out asynchronously, often referred to as 'start–stop' operation. This requires an additional element (the 'start' pulse) preceding the data elements, the receiving mechanism deriving a time reference from the arrival of its leading edge. The limitations of the mechanical teleprinter require that when a character has been printed the mechanism shall be allowed to come to rest before the following character is initiated and a time must be allowed in the signal stream for this decelleration (usually 1.5 elements 'stop pulse'). These two requirements led to the adoption of the standard 7.5 unit CCITT No. 2 (ITA-2 or Murray) code which has been in almost universal use for many years. There are two major drawbacks to the use of this code in the signalling path. First, one-third of the signal power (2.5 elements from a total of 7.5) conveys no data and is effectively wasted giving a 1.8dB loss of effective signal-to-noise ratio. Secondly, if detected by a start-stop mechanism, corruption of the first element will not only cause a character error but in the case of a continuous data stream can throw the printer out of synchronism and several subsequent characters may be corrupted. The increase in character errors owing to the defects of start–stop working over a 5-element synchronous system can be as high as 35:1 [10].

With the low cost of complex electronics resulting from modern technology there is no reason why these limitations of terminal telegraphy equipment should be applied to the signalling channel itself, and synchronous detection is becoming universal. It is in any case essential for operation on poor signals. Methods of converting an asynchronous data stream to synchronous are discussed in Chapter 10. However, a synchronous detection system usually requires a separate synchronising process in order to adjust the detection system to be in the correct time relationship with the incoming signal.

In a matched filter detection system the transition between two successive elements should ideally not occur during the sampling period and since only T/T_e of the transmitted power is utilised, the more accurately the receiving cycle is timed with reference to the signal frequency transitions, the higher the energy efficiency which may be attained.

The maximum achievable efficiency is limited by one or more of the following considerations:

(a) The accuracy with which the receiving system can identify in advance the instant at which the next element should begin.

(b) The variation in the actual time of arrival of an element from the expected, owing either to variations in the path delay of the signal or to instabilities in the send or receiving timing system.

(c) The length of time required to carry out the operations of comparison of the filter outputs, decision making, and dumping the filter energy (the 'dead time' in Fig. 2.4).

It is obvious that a system which depended on reliable identification of each signal transition before initiating the next sampling cycle would be hopelessly inefficient and therefore MFSK systems are essentially synchronous in operation, the element frequency $1/T_e$ being accurately maintained in the sending equipment and the cycle of the reception process also being accurately maintained at the same nominal frequency. A synchronising system is then required which adjusts the phase of the receiving cycle so that (in the ideal case) a signal transition does not occur in any sample period.

The effects of dead time, guard time and synchronising errors are discussed in detail in Sections 3.6 to 3.9 in which it is concluded that it is often possible to eliminate unnecessary guard time and to make the tone frequency separation equal to the inverse of the element period. The sampling time for which the switches are open is then slightly shorter than the theoretical ideal by the length of the dead time, but if this is a small fraction, its effect on the orthogonality of the filter response is acceptable. All MFSK systems designed by the author to date have used this relationship. This arrangement gives the absolute minimum of wasted signal energy, together with a considerable simplification of electronics. For initial analysis purposes, therefore, the process of comparison and dumping is assumed instantaneous. The function of the synchronising system is then to adjust the local receiving clock system so that this process coincides as accurately as possible with the effective instant of transition between one signal element and the next. The limitations on the accuracy of this process and various methods of achieving it are discussed in Chapter 8, and the effects of errors in Section 3.7.

2.8 References

1 SHANNON, C. E.: 'A mathematical theory of communication', *Bell Syst. Tech. J.*, July and Oct 1948, **27**, pp. 379–424 and 623–657

2 SLEPIAN, D.: 'Bounds on communication', *Bell Syst. Tech. J.*, May 1963, **42**, No. 3, pp. 681–708

3 VITERBI, A. J.: *Principles of Coherent Communication* (McGraw-Hill, 1966)

4 SANDERS, R. W.: 'Communication efficiency of several communication systems', *Proc. IRE*, April, pp. 575–588

5 SCHNEIDER, : 'Data transmission with FSK permutation modulation', *Bell Syst. Tech. J.*, July–Aug 1968, **47**, No. 6, p. 1131ff

6 CCIR 'Classification and designation of emissions', Kyoto, 1978, **I**, Rec. 507, p. 237ff

7 KOTEL'NIKOV, V. A.: *The Theory of Optimum Noise Immunity* (McGraw-Hill, 1959)

8 JORDAN, D. B., GREENBERG, H., ELDREDGE, E. E. and SERNIUK, W.: 'Multiple frequency shift teletype systems', *Proc. IRE*, November 1955, pp. 1647–1658

9 SCHWARTZ, M., BENNETT, W. R. and STEIN, S.: *Communication Systems and Techniques* (McGraw-Hill, Inter-University Electronics Series, vol. 4, 1966)

10 CCIR: 'Prediction of the performance of telegraph systems in terms of bandwidth and signal-to-noise ratio in complete systems', Kyoto, 1978, **III**, Rep. 195, p. 120

Chapter 3

Theoretical analysis

3.1 Frequency Response of a Matched Filter

Various implementations of a matched filter for a tone element are analysed in Chapter 9, but we shall consider initially the *LCR* circuit in Fig. 2.3 to which is suddenly applied at time $t = 0$ a signal current $I \sin \omega t$. (Mathematically speaking, this is identical with opening the switch while the signal is applied). The voltage across the circuit is given by:

$$C\dot{v} + \frac{v}{R} + \frac{1}{L}\int v \, dt = I \sin \omega t \tag{3.1}$$

The complete solution of this equation is of the form:

$$v(t) = \exp(-at) A \sin(\omega_d t + B) + bI \sin(\omega t + \phi) \tag{3.2}$$

It is unnecessary to evaluate here the circuit constants a and b, nor the constants A and B which depend on the initial conditions, but it should be noted that the assumption that the signal sine wave is at zero when $t = 0$ considerably simplifies the analysis with negligible loss of accuracy.

It can be seen that the response consists of two components:

(a) A 'transient component', consisting of a sinusoidal waveform of frequency ω_d, (the parallel resonant frequency of the tuned circuit), initially of amplitude A and decaying with time constant $1/a$.

(b) A 'steady-state component' consisting of a sine wave at the frequency of the input signal and amplitude bI, where b is a 'gain' parameter.

The responses of Fig, 2.3 may then be qualitatively explained. When t is very small, the two components are equal and in anti-phase, so that the response increases from zero as the transient component dies away. If the signal frequency is off-tune the two components beat together to give an oscillatory envelope at the difference between the signal and resonant frequencies. When the losses are reduced to zero (R increased to infinity), then $a=0$, the 'transient' response is maintained indefinitely and the damped resonant frequency ω_d becomes the lossless resonant frequency ω_0 given by:

$$\omega_0 = (LC)^{-\frac{1}{2}}$$

Under these conditions eqn. 3.2 becomes:

$$v(t) = \frac{I\omega}{C} \frac{\cos \omega_0 t - \cos \omega t}{(\omega_0 + \omega) \quad (\omega_0 - \omega)} \qquad (3.3)$$

Now if $(\omega_0 - \omega) = \Delta\omega$ and $\Delta\omega \ll \omega$

$$v(t) = \frac{It}{2C} \frac{\sin (\Delta\omega t/2)}{\Delta\omega t/2} \sin \{(\omega_0 + \omega)\, t/2\} \qquad (3.4)$$

This is the equation plotted in Fig. 2.3b. The last term is a sine wave at the mean of the resonant frequency and the signal frequency so the envelope of the output is:

$$V(\omega,t) = \frac{It}{2C} \left| \frac{\sin (\Delta\omega t/2)}{\Delta\omega t/2} \right| \qquad (3.5)$$

The relationship between V and ω for a specific time $t=T$ gives the effective frequency response of the filter at the comparison time.
Then:

$$V(f) = \frac{IT}{2C} \left| \frac{\sin \pi \, \Delta f T}{\pi \, \Delta f T} \right|$$

or (3.6)

$$V(n) = V_0 \left| \frac{\sin n\pi}{n\pi} \right|$$

where $\Delta f = \Delta\omega/2\pi$, $n = \Delta fT$, and V_0 = output at time T of the filter at the frequency of the signal.

This equation is plotted in Fig. 3.1. It demonstrates the orthogonal relationship defined in Section 12.4 since $V_n = V_0$ for $n=0$, and zero for any other integral value of n (positive or negative). It is very important to note that a filter is matched to a signal of specific frequency and *of specific duration*. The true 'output' of a matched filter consists of an instantaneous sample at a specific time of the envelope of its response, and the output in any other sense is irrelevant except when considering details of design. The timing and switching process is therefore an integral and indispensable part of the matched filter process. In this sense the term 'matched filter' may be misleading and 'matched detector' would be preferable, but the former term is retained as being generally accepted.

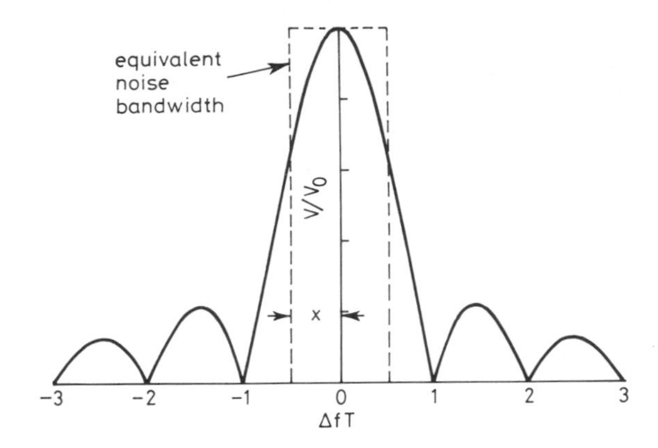

Fig. 3.1 *Frequency response of a matched filter output (at time T)*

3.2 Noise Bandwidth of a Matched Filter

A primary factor in the performance of a signalling system is its behaviour when noise is added to the signal. It is necessary therefore to consider the output of a matched filter when energised by white Gaussian noise. This is expressed in terms of an 'equivalent noise bandwidth' (ENB) defined as the bandwidth of an ideal bandpass filter which has the same response to the signal frequency, and the same RMS output (over a large number of samples) when excited by white noise alone (both outputs being measured *at time* T). From this definition it is evident that the ENB is given by:

$$B_n = \frac{1}{V_0^2}\int_{-\infty}^{\infty} [V(f)]^2 \, \mathrm{d}f$$

$$= \int_{-\infty}^{\infty} = \left[\frac{\sin x}{x}\right]^2 \frac{dx}{\pi T} \quad \text{(where } x = \pi\,\Delta f T\text{)}$$

$$= 1/T \tag{3.7}$$

The RMS of a large number of output samples of a matched filter energised by white noise is therefore the same as that of the envelope of the output of an ideal bandpass filter having the same response to the signal and a bandwidth 1/T where T is the sample period. (As indicated in Fig. 3.1).

Note again that this refers to samples of the output envelope at time T and not to the bandwidth of the resonant circuit itself, which is zero. From eqn. 3.6 for a fixed input signal level, V_0 is proportional to T, while from eqn. 3.7, for a fixed input noise density (input noise power per Hz of bandwidth), the output noise *power* is proportional to T. It follows therefore that for a given input signal-to-noise ratio, *the output signal-to-noise ratio of a matched filter is proportional to* $\sqrt{T}$.

3.3 Error Probability with Added Noise

Consider the MFSK system of Fig. 2.1 sending a sine wave signal of unity RMS amplitude at one of the filter frequencies, with white Gaussian noise voltage $N(t)$ added in the signalling channel, where $N(t)$ has a power density of $\bar{N}^2$ watts per Hz of bandwidth. ($\bar{N}^2$ is then the ratio (noise power per Hz bandwidth) / (signal power)).

Then

$$V(t) = \sqrt{2} \sin \omega t + N(t) \tag{3.8}$$

Then after a time T the output of the filter at the signal frequency will consist of the envelope of a waveform $V_s(t)$ containing two components, one due to the signal and the other a noise component equivalent to that from an ideal filter of bandwidth $1/T$. Assuming a unity gain constant:

$$V_s(t) = \sqrt{2} \sin \omega \; T + \bar{N}/\sqrt{T} \text{ volts} \tag{3.9}$$

Since the other filter outputs at time T are orthogonal to the signal, the noise component only is present, so the envelope of the output from any of the other $(M-1)$ filters is that of a noise source of RMS voltage:

$$V_N(t) = \bar{N}/\sqrt{T} \text{ volts} \tag{3.10}$$

The theoretical probability of an error is the probability that the envelope of $V_s(t)$ is less than one or more of $(M-1)$ independent samples of the envelope of $V_N(t)$. This problem has been treated by Ward [1] and an abbreviated analysis is as follows:

The probability density $p(v)\,dv$ of a voltage is the probability that a sample of the voltage will lie between v and $(v + dv)$. Rice [2] has shown that if the voltage is the envelope of a steady sine wave of power W_s added to a Gaussian noise voltage of power W_N then:

$$p(V_s)\,dv = \frac{v}{W_N}\exp-\frac{W_s+v^2}{2W_N}\,.I_0\frac{vW_s}{W_N}\,dv \qquad (3.11)$$

where $I_0\,(x)$ is the Bessel function of imaginary argument and order 0. For the envelope of noise alone the probability density is:

$$p(V_N)\,dv = v\exp(-v^2/2W_N)\,dv \qquad (3.12)$$

The cumulative probability distribution $P(v)$ is the probability that a sample of the voltage is less than v and it is evident that:

$$P(v) = \int_0^v p(v)\,dv$$

which for the envelope of noise alone gives:

$$P(V_N) = 1-\exp(-v^2/2W_N) \qquad (3.13)$$

Since there are $(M-1)$ filters containing noise alone, the probability that at any single comparison time, the output of all such filters is less than v is:

$$P'(V_N) = (P(V_N))^{M-1}$$

The probability that one or more of the noise samples exceeds the voltage v is then given by $\{1-P'(V_N)\}$. So the total probability of error in a single selection from M filter outputs is:

$$p_e = (1-P(V_N)^{M-1})\;p(V_s)dv$$

$$= \int_0 \left\{1-\left[1\text{-exp}\,\frac{-v^2}{2W_N}\right]^{M-1}\right\}\frac{v}{W_N}\exp-\frac{W_s+v^2}{2W_N}\,.I_0\frac{v\sqrt{W_s}}{W_N}\,dv \qquad (3.14)$$

Ward has shown that this reduces to

$$p_e = \sum_{r=1}^{M-1}\frac{(-1)^{(r+1)}\,(M-1)!}{(r+1)!\,(M-r-1)!}\exp\left(\frac{-r}{r+1}\,\frac{T}{\bar{N}^2}\right) \qquad (3.15)$$

Note that $(T/\bar{N}^2)$ is the signal-to-noise power ratio in the signal filter at the end of the element. A useful rule-of-thumb is that the breakdown (≃ 1% errors) of most MFSK systems occurs when this ratio is about 10:1 (+ 10 dB).

The probability of error in identifying the correct filter at the end of a single element is p_e, but if C elements convey a single character (symbol or byte) of

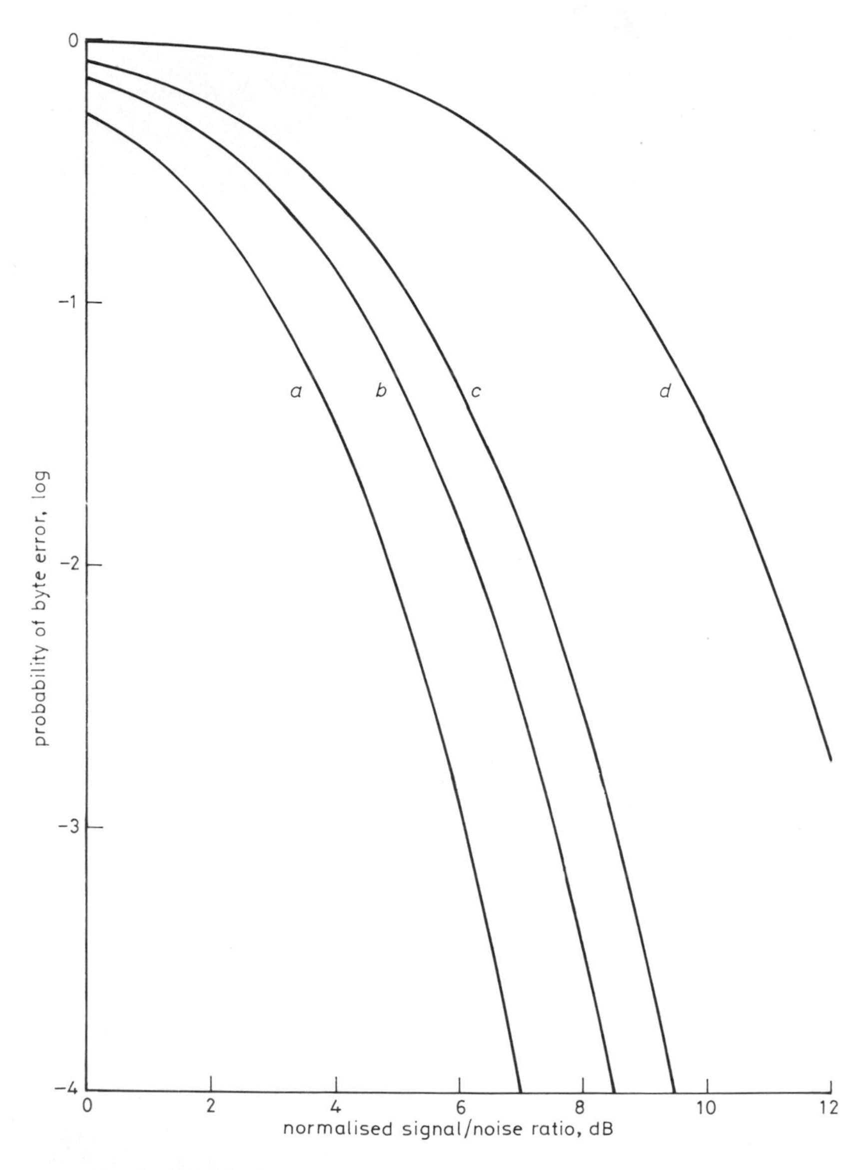

Fig. 3.2 *Byte error rate of four alternative MFSK systems*
(For parameters see Fig. 2.6)
Non-fading signal; non-diversity reception

information, an error in any one or more of C elements will cause a character error so that:

$$p_C = 1 - (1 - p_e)^C \tag{3.16}$$

$$\simeq Cp_e \text{ for small values of } p_e$$

The character error rate for any MFSK system in white Gaussian noise may therefore be computed from eqns. 3.15 and 3.16 but note that the former computation requires care, since it is a summation of terms of alternate signs, and while some values are very large (32! exceeds 10^{34}), the final sum is the element error rate, which may be of interest below 10^{-4}.

The performance of the systems illustrated in Fig. 2.6 are shown in Fig. 3.2. It can be seen that the conclusions of Slepian and Viterbi quoted earlier are confirmed, in that the performance improves as M is increased, but that the rate of improvement is reduced at higher values of M. Note also that the assumed values of T are those given in Table 2.1, giving the same overall data rate, and that the horizontal ordinate is the 'normalised signal-to-noise ratio' defined as:

$$R_o = \frac{\text{signal energy per bit}}{\text{noise power per Hz bandwidth.}}$$

$$= \frac{T}{\bar{N}^2} \frac{1}{\log_2 M}$$

3.4 Bandwidth and Spectra of MFSK

3.4.1 Necessary Bandwidth

The 'necessary bandwidth' of a receiving system is defined [3] as that which 'is just sufficient to ensure transmission of information at the rate and quality required under specified conditions'. The frequency spectrum of a single element of signal may be calculated by Fourier integral methods, but the calculations are considerably simplified by assuming that the signal is periodic, consisting of a sine wave modulated 'on' and 'off' for alternate elements of duration T. The frequency spectrum is then given by:

$$v(t) = \left[\frac{1}{2} + \frac{2}{\pi} \sum_{n=1}^{\infty} \frac{1}{n} \sin n\pi\, t/T\right] \sin \omega t$$

$$= \frac{1}{2}\left[\sin \omega t + \frac{2}{\pi} \sin\left(\omega \pm \frac{\pi}{T}\right)t + \frac{2}{3\pi} \sin\left(\omega \pm \frac{3\pi}{T}\right)t \ldots\right] \tag{3.17}$$

An ideal bandpass filter restricting the bandwidth to the carrier and first pair of sidebands only would accept approximately 90% of the transmitted power, while inclusion of the second pair increases this to 95%. This is an idealised concept and considerations of group delay (see Section 8.1) will normally necessitate a wider margin. It is therefore reasonable to suggest that *the minimum necessary bandwidth for the input signal to a single matched filter extends about twice the inverse of the integration period on either side of the filter frequency.*

3.4.2 Frequency Spectrum

The frequency spectrum of an MFSK signal modulated by data will of course be a function of the data being carried, but a worst-case analysis may be made by assuming that the system is keying alternately between the maximum and minimum frequencies. The formula derived for FSK keying with zero build-up time may then be used. An empirical formula [4] is:

$$A_X = \frac{2}{\pi \mathrm{m}(x^2-1)} \tag{3.18}$$

where m = modulation index
= (total frequency change)/(modulation rate in bauds)
x = (frequency from centre)/(half total frequency change)
A_x = amplitude of component at frequency x.

These parameters can be converted into those used in the analysis of MFSK in Chapter 2. Since the spectrum is symmetrical, it is convenient to express it in terms of the amplitude A_n of a component at a frequency (n/T) *above the higher keyed tone*, which is itself at a frequency $(M-1)/2T$ above the centre of the spectrum. The equation is then modified to:

$$A_n = \frac{2T_e}{T(M-1)} \; \frac{1}{(x^2-1)} \tag{3.19}$$

where $x = \{1+2n/(M-1)\}$.

This spectrum is plotted in Fig. 3.3 for three values of M and assuming $T_e = T$ (see Section 3.8). It can be seen that even for $M=32$ the component at four tone intervals away from the extreme tones is almost 30 dB down on the unmodulated carrier.

3.4.3 Occupied Bandwidth

The occupied bandwidth of a signal is defined [3] as the bandwidth containing 99% of the total mean transmitted signal power, 0.5% being excluded above its top limit, and 0.5% below its bottom limit. Eqn. 3.19 may be used to calculate the power excluded by an ideal bandpass filter stopping all components at $\pm$n and beyond. This is plotted in Fig. 3.4 and it can be seen that for $M=32$ the

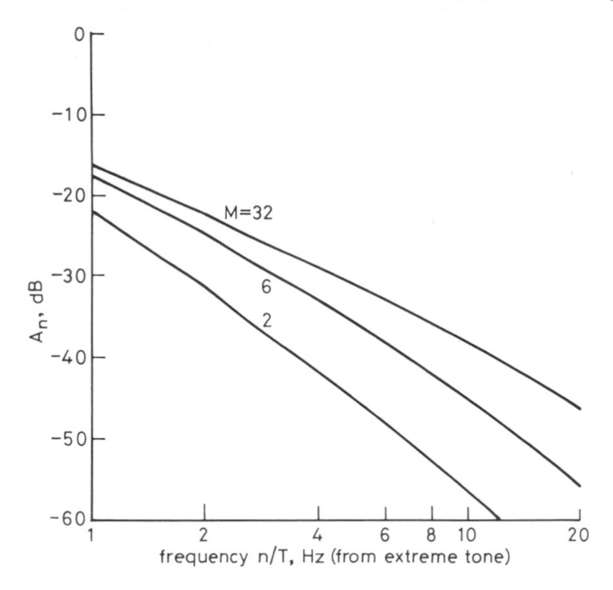

Fig. 3.3 *Frequency spectra of MFSK systems*

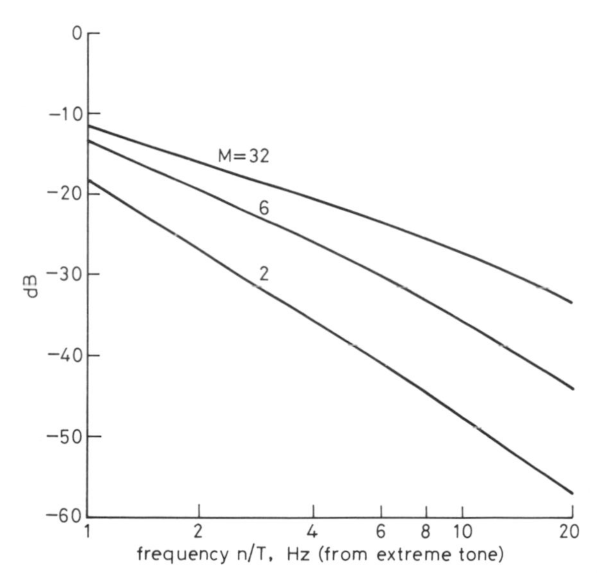

Fig. 3.4 *Signal power rejected by an ideal bandpass filter with cut-off at* $\pm n/T$ *Hz from extreme tones*

occupied bandwidth extends to about 3.5 tone intervals above and below the extreme keyed frequencies, while for $M=6$ it is little more than 2 tone intervals. When it is considered that this is a worst-case calculation and for large values of M keyed with random data the spectral energy will be concentrated nearer to the centre of the transmission band, it is not unreasonable to suggest that a good rule of thumb covering most practical cases is that the occupied bandwidth extends from about 2 tone intervals below the lowest tone frequency to 2 tone intervals above the highest, i.e. it covers the necessary bandwidth of the extreme filters and the band in between. This approximate equivalence of the two bandwidths implies excellent bandwidth efficiency and corresponds to the definition of a 'perfect' emission [3].

3.4.4 Normalised Bandwidth

It is now possible to calculate a 'normalised bandwidth', 'specific bandwidth' or 'bandwidth occupancy' defined as:

$$B_o = \frac{\text{occupied bandwidth}}{\text{data rate}} \text{ Hz/bit/s}$$

Let the occupied bandwidth extend a *total* of (G/T) Hz (i.e. G tone intervals) beyond the extreme tones. Then the occupied bandwidth is $(M-1+G)/T$ Hz, and since the data rate (from eqn. 2.6) is $(\log_2 M)/T_e$ bits/s:

$$B_o = \frac{(M-1+G)T_e}{T\log_2 M} \text{ Hz/bit/s} \tag{3.20}$$

The curves of Fig. 3.5 show B_o against M for various values of G (plotted as continuous curves for convenience and again assuming $T_e = T$). An appropriate

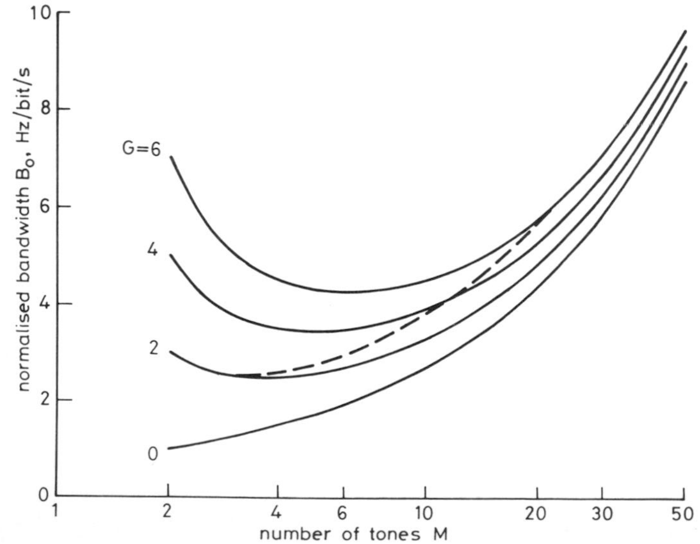

Fig. 3.5 *Normalised bandwidth of MFSK systems*
- - - - occupied bandwidth (approximately)

value of G for the occupied bandwidth (defined above) can be read from the curves of Fig. 3.4, but because of the discrete nature of the spectrum the resulting curve is discontinuous and the dotted line in the figure is an approximation. It can be seen that the minimum normalised bandwidth of about 3.0 is obtained for $M=3$ to 6, and for values of M greater than about 20 the normalised bandwidth increases proportionally to M.

These calculations apply to a single channel of MFSK but if several channels are frequency-division multiplexed the normalised bandwidth is reduced. It would seem practicable to operate a number of MFSK channels in fdm with the frequency spacing between each channel equal to the spacing between tones. If all element timings are synchronous, the signals in each channel are orthogonal to each other and there is no interchannel interference. The guard band allowance then applies at the extremes of the spectrum only. For example, 20 channels of 6-tone MFSK would give a total normalised bandwidth of

$$\frac{(6 \times 20 - 1 + 4)}{20 \log_2 6} = 2.4 \text{ Hz/bit/s}$$

This possibility is discussed in more detail in Sections 7.6 and 12.4.

3.5 Frequency Error

Before proceeding to deduce optimum values of parameters for any particular application it is necessary to consider the effects of divergencies in the practical system from the ideal conditions discussed in Chapter 2. If the signal is noise-free the allowable divergencies can be quite large, but this is an impracticable assumption, and in fact the simplest approach to estimating such effects is to assume a noisy signal giving a low but finite error rate and then to calculate the change in signal-to-noise ratio required to maintain the same error rate when the unwanted effect is introduced.

The initial problem to be approached in this way is to calculate the effect of a small constant frequency error in the received signal, such as may arise in the design or adjustment of the equipment, or in an associated frequency standard. The extension of these conclusions to dynamic variations arising in the signal path will be considered in Chapter 6.

The frequency response of a matched filter is given in Fig. 3.1. If the signal frequency is offset from its correct value this will have two effects.

It will reduce the output of the 'correct' filter (i.e. the one corresponding to the signal frequency).

It will generate a spurious output in each of the 'incorrect' filters.

From eqn. 3.6, if there is a frequency error in the signal of (a/T)Hz (i.e. an offset of 'a' times the frequency interval between filters, where $a<0.5$), then the response V_n of a filter separated 'n' tone intervals from the correct filter in the same direction as the error is:

$$V_n = V_0 \left| \frac{\sin(n-a)\pi}{(n-a)\pi} \right| \tag{3.21}$$

With no frequency error $a=0$, and if noise causes an incorrect filter to be selected, it is because on that occasion a negative peak (minimum) of noise on the correct filter $(n=0)$ reduced its output below a positive peak of noise on an incorrect filter output. It can be shown (and it is intuitively evident) that the greatest probability of this happening is when both noise amplitudes are approximately equal to $V_0/2$. Now consider a similar event with a frequency error which reduces the output of the correct filter to V'_0. It is evident that the amplitude of noise now required to cause a similar probability of a decision error is approximately $(V'_0 - V_n)/2$.

Therefore, assume no frequency error, and noise such as to give a specific probability of selection error p *by selecting a specific filter at frequency n.* Now if the frequency error is applied, the error probability will increase. To restore the error rate to p again will require the noise input to be reduced by a factor:

$$\begin{aligned} r_f &= \frac{(V'_0 - V_n)/2}{V_0/2} \\ &= \frac{\sin a\pi}{a\pi} - \left| \frac{\sin (n-a)\pi}{(n-a)\pi} \right| \\ &= 1 - \left| \frac{a}{n-a} \right| \frac{\sin a\pi}{a\pi} \end{aligned} \tag{3.22}$$

Thus, although it would be difficult to calculate the increase in error rate produced by a given frequency error, it is simple to calculate the equivalent deterioration in signal-to-noise ratio for each 'incorrect' filter and these are shown in Fig. 3.6. As would be expected, the worst case is for $n = +1$, i.e. the filter most likely to be selected in error is the adjacent on the same side as the frequency error. In this case the deterioration will approach infinity for $a=0.5$, since an error of more than half a tone interval will cause the adjacent filter to be selected even in the absence of noise. This curve may be used as a rough indication of the total deterioration but can be shown to be pessimistic, and more detailed analysis suggests that the true curve for total effective signal-to-noise deterioration will lie above the curve for $n = +1$ by about 0.7 dB at $a=0.2$, i.e. will approximate to the curve for $n = -1$. As a rule-of-thumb the effective signal-to-noise ratio will therefore deteriorate about 1 dB for each 10% increase in a for small errors. Measurements on a practical system gave approximately 1.5 dB deterioration for 20% frequency error.

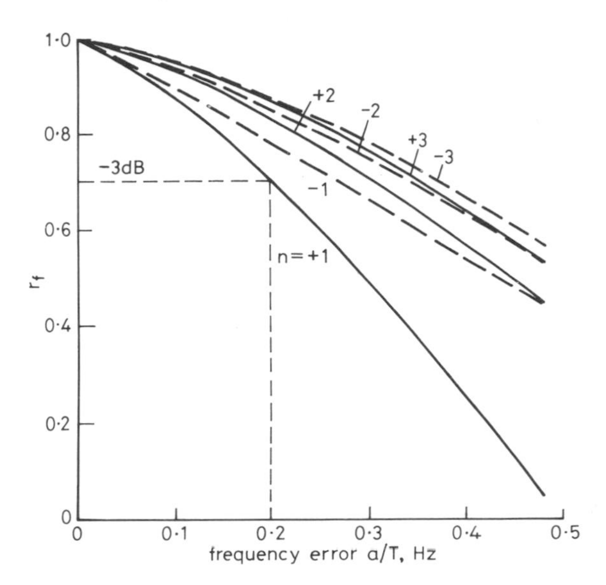

Fig. 3.6 *Losses with frequency error*

3.6 Dead Time Error

The filter responses are orthogonal if the sampling time T (i.e. the time for which the switches are open, see Fig. 2.4) is equal to the inverse of the frequency interval between adjacent filters. Assume that this condition is not met, and that the sample time is reduced to $(1-b)T$. where b is a small factor referred to as the 'dead time' (the connection between this and the dead time of Fig. 2.4 will be explained later). Then from eqn. 3.5 the response of a filter separated by (n/T) Hz from the correct one in either direction is

$$V_n = V_0\,(1-b)\left|\frac{\sin n\pi(1-b)}{n\pi\,(1-b)}\right| \tag{3.23}$$

Following the same approach as in Section 3.5, in the absence of dead time $b = 0$ and the difference between the correct and any incorrect filter equals V_0. If b is finite the reduction in noise level required to restore the same error probability owing to wrong selection of any filter n is given by:

$$r_d = (1-b)\left(1-\left|\frac{\sin n\pi\,(1-b)}{n\pi\,(1-b)}\right|\right) \tag{3.24}$$

$$\simeq (1-2b)$$

This relationship is shown in Fig. 3.7 and it can be seen that it may be assumed that the equivalent signal-to-noise ratio will deteriorate about 1 dB for 5% of dead time.

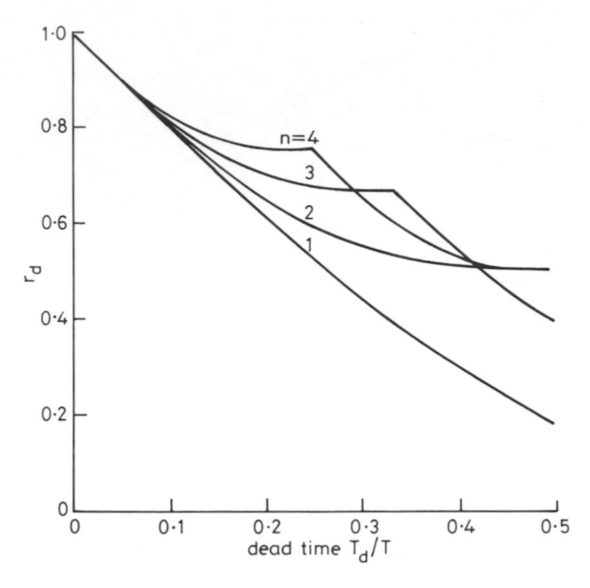

Fig. 3.7 *Losses with dead time*

3.7 Synchronising Error

A synchronising error occurs if the relative timing of the demodulation process and the incoming element, as shown in Fig. 2.4, is wrong, so that the transition between one element and the next occurs during the sample time T, the matched filters being subject to the wrong frequency for a period cT and the correct frequency for a period $T(1-c)$, where $c<0.5$. It can be shown that it is immaterial whether the wrong signal occurs at the beginning or at the end of the sample period, and in the following discussion it is assumed to be at the end. This is a more complex case than the dead time, since during the error period a spurious signal from the succeeding element is being fed into the filters. The effects will therefore depend upon the separation between the two signal frequencies, and whether or not there is a phase transient at the frequency transition. The latter effect is to be avoided (if only because it creates unnecessary out of band interference products) and therefore the assumption may be made that the frequency transition is phase-continuous. After the frequency transition, therefore, the signal will begin to deviate in phase at a rate proportional to the frequency change. The largest possible frequency change is $(M-1)T$ so that at a time $0.25T(M-1)$, the phase error will be less than 90°. The effects on the orthogonality of the filter responses will be negligible for at least this period, if not considerably longer. In effect, the wrong frequency is initially a good approximation to the correct signal. The relationship derived for the dead time may therefore be used as a pessimistic upper limit of deterioration due to synchronising error.

The conclusion that the effects of synchronising error depend upon the frequency change was supported by measurements on a practical system which has a nominal sample time of 50 ms which actually included a dead time of less than 2 ms (4%) as discussed below. When keying frequencies 500/520 Hz a synchronising error of ±20% of the sample period gave an error rate deterioration equivalent to an increase in noise level of less than 1 dB. When keying 420/620 Hz a similar deterioration was produced by a synchronising error of about ±8% (approaching the limit set by the calculations on dead time alone).

Under time-varying synchronising error conditions such as may be experienced with multipath reception (see Section 4.7), the situation is even more complex, being confused by the accompanying fading phenomena. Measurements made under these conditions (shown in Fig. 7.5 and discussed in Section 7.7) roughly support the above analysis.

3.8 Guard Time

Referring to Figs. 2.4, 3.8*a* and 3.9*a*, it is now possible to estimate the losses due to the various parameters in the receive timing cycle. Starting from an idealised reference system in which the guard time, dead time and synchronising error are all zero and $T = T_e$, then the introduction of a synchronising error will cause a deterioration not worse than that given by the dead time eqn. 3.24, and so may be expressed as $r_s = 1 - 2bK$ where $K \leqslant 1$. This is shown in the solid lines of Fig. 3.9*a* for three values of K.

From Section 3.2 the introduction of a guard time will, in the absence of a synchronising error, reduce the signal-to-noise ratio at the correct filter output by a factor:

$$r_g = (T/T_e)^{\frac{1}{2}} \simeq 1 - T_g/2T_e \qquad (3.25)$$

Introduction of a dead time less than the guard time will now have no effect, and a synchronising error less than half the guard time will also have no effect. Synchronising errors in excess of this will reduce the signal-to-noise ratio as before, so that the total losses with synchronising error are then given by:

$$r'_g = (1 - T_g/2T_e)\,\{1 - K(2t_e - T_g)/T\} \qquad (3.26)$$

This is shown by the dotted lines in Fig. 3.9*a* for a guard time of 30% ($T_g = 0.3T$). It can be seen that for small synchronising errors or small frequency changes the losses with a guard time are worse than those without.

It is reasonable to consider an alternative arrangement in which there is no guard time and the timing cycle is then revised as Fig. 3.8*b*, where nominally $T =$

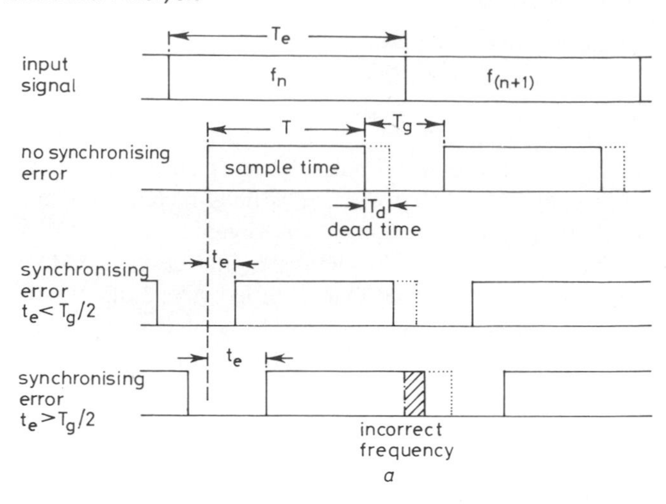

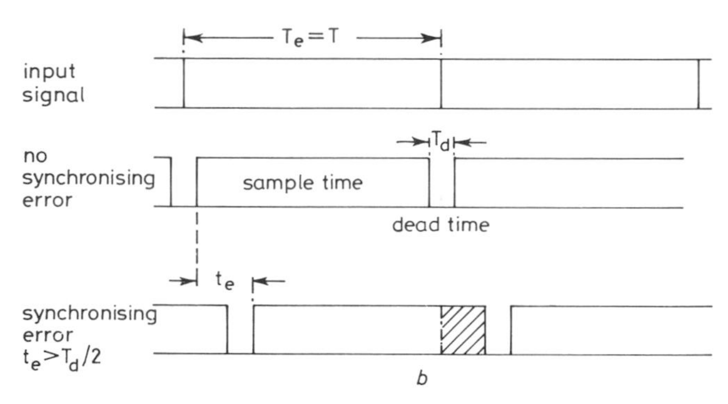

Fig. 3.8 *Effect of synchronising error with alternative timing systems*
a Receive system timing with guard time
b Receive system timing with no guard time

T_e but in fact the matched filter switches are open for a period $(T - T_d)$ where T_d is the minimum dead time necessary to discharge the filters. Then in the absence of synchronising error the effective signal losses are given by eqn. 3.24 with $b = T_d/T$, and for synchronising errors exceeding half the dead time:

$$r'_d = (1 - 2T_d/T)\{1 - K(2t_e - T_d)/T\} \tag{3.27}$$

The curves for a dead time of 5% are then as shown dotted in Fig. 3.9*b*. Comparing the two situations, it would seem that the argument in favour of a guard time is by no means as clear cut as may at first be supposed, and the provision of a guard time as a protection against synchronising and multipath errors, as is done in many modern binary telegraphy systems (see Section 7.4.1), is of dubious advantage under the conditions analysed here. A further factor is that the provision of a guard time slows down the data rate, increasing the normalised bandwidth according to eqn. 3.20.

These disadvantages are such that in all the MFSK systems designed by the author to date, no guard time has been incorporated. In the most recent of these systems the dead time is less than 3% and could be reduced if required. However, consideration of the synchronising errors likely to be attained in practice suggest that further reduction of the dead time would not materially improve the overall performance.

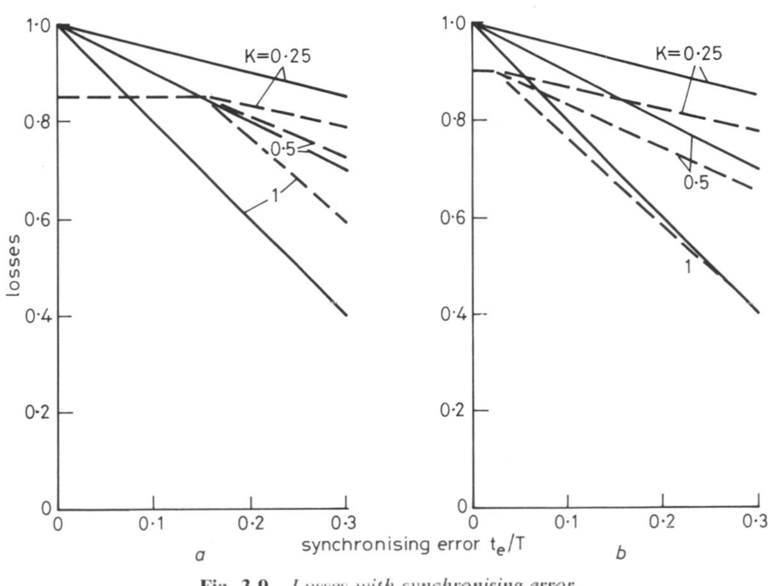

Fig. 3.9 *Losses with synchronising error*
a With guard time
- - - - $T_g = 0.3T$
b No guard time
- - - - $T_d = 0.05T$

3.9 Basic Conclusions

From the work so far, it is possible to draw a number of general guide lines on the design of an MFSK system.

1 Provided that the mean absolute time delay of the communication path is relatively stable in the long term (which is usually the case), a good communication system can and should exploit to the full the ease with which high stability tone and clock frequencies may be generated, and therefore true synchronous operation is essential. The instabilities and errors in frequency and timing within the equipment should be negligible compared to those in the communication channel.

2 The use of a guard time as a protection against time dispersal effects is not always effective, and any benefits should be weighed against the waste of transmitter energy. In assessing the effects of time dispersal, full account should be taken of the characteristics of the synchronising system.

3 In the design of a basic system as described above, the key parameter is the duration of an element and this must be a compromise based on the consideration of three factors:

(a) The frequency dispersion of the communication medium. (Phase and frequency instability, fading etc., see Sections 4.6 and 4.8).

(b) The time dispersion of the communication medium (multipath, see Section 4.7).

(c) The signal-to-noise performance required.

If bandwidth is not a limitation then the element length should be the maximum possible consistent with (a) as this will give the best performance against (b) and (c).

4 If bandwidth must be reduced to the minimum then the optimum number of tones is between 4 and 10 provided that this results in an element length which is acceptable for (a) and (b). This procedure results in a single channel of fixed data rate giving the minimum bandwidth occupancy.

5 If bandwidth is less critical, then the number of tones may be increased to give better performance in noise, to give the data rate required, or to an economic maximum, (but values of M higher than about 20–30 give a proportional increase in bandwidth with very little improvement in performance against noise).

6 Major changes in the characteristics of the system may be achieved by variations on the basic system (as described in Chapter 12), but since this is so near to the limits of the theoretically attainable performance in most respects, it is almost inevitable that any variations producing an improvement in performance in one respect will produce a deterioration in another.

3.10 References

1 WARD, H.: 'On the probability of error for largest of selection', (Letter) *Proc. Inst. Radio Eng.*, Jan 1962, **50**, p. 93
2 RICE, S. O.: 'Statistical properties of a sine wave plus random noise', *Bell Syst. Tech. J.*, 1948, **27**, pp. 109–157
3 CCIR: 'Spectra and bandwidth of emissions', Kyoto, 1978, Rec. 328–4, **I**, p. 306ff
4 CCIR: 'Bandwidth of radiotelegraph emissions types A1 and F1', Kyoto, 1978, Report 179–1, **I**, p. 263ff

4

Characteristics of the H.F. Communication Channel

4.1 Model of a communication link

The Piccolo communication systems referred to in Chapter 1 were designed to operate on the network of point-to-point h.f. telegraphy links operated by the Communications Division of the UK Foreign and Commonwealth Office and discussed in Chapter 6. This type of operation encounters most of the problems of poor signal h.f. telegraphy, and will frequently be used as a working example in this book. No attempt will be made here to explain or analyse in detail the complexities of ionospheric propagation but instead the various effects of the communication path on the signal will be simplified (and indeed over-simplified by many standards) and separated as far as possible into mathematically defined parameters applicable to the model illustrated in Fig. 2.1. The effects considered are:-

Interfering signals;
Added white Gaussian noise;
Added non-Gaussian (impulse or 'static') noise;
Slow flat fading;
Multipath propagation (and selective fading);
Doppler shift;
Rapid fading.

It will be noted that this list includes many of the effects found in other types of communication channel, although of course the actual values of the parameters may be grossly different. The general line of approach indicated here will therefore be useful for analysis of channels other than h.f., and one or two examples of different communication media are discussed in Chapter 12.

4.2 Interference

This term is normally applied to the reception, in addition to the wanted signal, of an unwanted man-made signal of any kind, whether or not that signal was deliberately generated for communication purposes. Such a classification embraces a wide field, but some of the common types are discussed here.

4.2.1 Tone signals

In many cases an interfering signal consists of an unmodulated carrier, which, particularly if the wanted signal is being detected by SSB methods, produces a single audio tone (often drifting in frequency across the signalling band). When the frequency coincides with one of the matched filters, and if the amplitude exceeds that of the wanted signal, then an incorrect filter may be selected. If the interfering signal is interrupted or modulated in any way (for example a Morse signal or one tone of an FSK signal) its effect is reduced. The matched filter effectively integrates the signal within its bandwidth over the sampling time, so that if the interference is only within that band for say 20% of the duration of a particular element its amplitude must be at least five times that of the signal before it will create an error.

4.2.2 Impulsive noise

Rotating electrical machinery (such as lift motors, vacuum cleaners etc.) can radiate trains of pulse interference at a pulse rate related to the speed of rotation of the motor. Fluorescent lights, thyristor control systems and similar devices can produce interference on harmonics of the mains frequency. The duration of the pulses from such sources is usually narrow compared with the repetition period, resulting in a very wide power spectrum with very little energy in any one component. Since the bandwidth of each matched filter in an MFSK system is so narrow, the effect on such a system is usually negligible even when the wanted signal as monitored on a loudspeaker or oscilloscope is apparently completely swamped by the interference.

However, such interference, when present at large amplitudes, may affect the signalling system through indirect effects in some non-linearity of the signal circuits or in the AGC system of the radio receiver. For instance, a sudden high-level pulse of interference can cause short term overloading of an early stage of the receiver, causing ‘blocking’ of the wanted signal during the impulse and a reduction in gain for a subsequent ‘hangover’ period. The narrow bandwidth of the i.f. filters may modify the waveform, and then the AGC system (which may not respond to the steep wavefront) will modify the receiver characteristics to an extent depending on the duration of the impulses or their mean repetition frequency. In an extreme case the audio output from the radio receiver may be defined more by the characteristics of the receiver itself than those of the incoming waveforms.

4.2.3 Speech and music

The spectra of radio telephony signals are also relatively wide and again the interference can be of very high amplitude compared with the signal before it has any appreciable effect on signalling accuracy. In some experiments carried out by the author for the British Broadcasting Corporation with an early model of Piccolo, the Piccolo signal was added to an audio signal of 'typical' programme material (including brass bands, electronic music, applauding crowds, sound effects and drama with a hysterical heroine) at various levels and the effect on the signalling accuracy of Piccolo and the audibility of the programme assessed. With the Piccolo signal in its original frequency band of 330–650 Hz, appreciable errors were produced only if the signal was less than -6 dB on peak programme level and then only on sustained notes in electronic music, the chimes of Big Ben etc. The Piccolo signal was then moved to 2950–3270 Hz and the level could then be reduced below -25 dB on peak programme level before errors were unacceptable. These were produced mainly by loud brass (trumpets) or sustained high notes on a guitar. At this level the programme was comfortably audible with the Piccolo signal little more than an irritating nuisance in the background. For these experiments the two signals were added and assessed without any filtering. It was concluded that if the signal was introduced into a filtered slot in the programme spectrum, at a frequency of 4 kHz or higher, the Piccolo level could be reduced to about -30 dB and it would then have little or no effect on signal quality (by short-wave relay standards).

4.3 Aerial and atmospheric noise

Thermal noise generated in the resistive component of the aerial source impedance or input impedance of a radio receiver is white (i.e. has a constant power density with frequency) and has a Gaussian probability distribution. However, noise from these sources is not normally the limiting factor in h.f. communication (except possibly at higher frequencies on quiet sites), being usually of lower amplitude than interference or atmospheric noise.

Atmospheric noise is mostly generated from thunderstorm activity (predominantly in the tropics) and within a few hundred miles of the storm may be received as discrete 'crashes' associated with specific discharges (often referred to as 'static'). These may be considered in the same way as man-made impulsive interference. At longer ranges the large number of such storm centres active at any one time, plus the dispersive effect of ionospheric propagation, result in a more or less continuous background of noise generally similar to thermal noise but with a strong impulsive content and varying with latitude, season, time of day etc. Mathematical analysis of system performance under such conditions is difficult and is in any case unnecessary in this book, where the aim is usually to deduce the broad optimum of a parameter value, or to compare the performances of two basically similar systems, and for such purposes the white Gaussian model for noise is adequate. It is therefore acceptable, and almost universal practice, to assume white Gaussian noise in calculations such as those in Sections 3.2 and 3.3

4.4 Characteristics of fading

An electromagnetic wave propagated in an ionised medium and entering obliquely a region of increased ionisation, has its direction of propagation deflected towards the region of lower ionisation. The various ionospheric layers which surround the earth at altitudes of 80–500 km provide this condition, so that a radio wave travelling upwards at an oblique angle is successively deflected towards the earth, and if certain limiting conditions are satisfied it will re-emerge from the underside of the layer to be returned to the earth's surface a distance away. This process is roughly similar in effect to a reflection from a solid surface and for simplicity will be referred to as such. However, the ionosphere is not homogenous or static, but contains regions of higher ionisation density which are constantly moving. The result is that, unlike a specular reflection, the received signal is the sum of a number of components, each of which has travelled a slightly different path and so has experienced a slightly different delay. These differential delays are continually varying so that components which are in phase at one instant may be in anti-phase some time later. It should be noted that on a frequency of 10 MHz a differential variation of 0.05 microseconds (or 15 metres of path) between two components is enough to cause such a change. The enhancement or cancellation of in-phase and out-of-phase components leads to fluctuations in amplitude and phase of the received signal, referred to as 'fading'. Mathematically, the problem of analysing the sum of a large number of sine waves of varying phase can be reduced to the familiar case of a two-dimensional 'random walk' and the result can be shown to have the same statistical characteristics as narrow-band Gaussian noise [1]. In particular, the envelope amplitude has a Rayleigh distribution, the power spectrum is Gaussian (or normal), the phase is randomly distributed over 2π and there is a correlation between the envelope amplitude and the 'instantaneous frequency' (see below).

Although the model of the ionosphere as described will produce these characteristics (generally referred to as 'Rayleigh fading'), other types of fading are possible [3]. For instance, if the receiving station is within the range of the direct ray (or 'ground wave') from the transmitter, the signal may consist of the sum of a Rayleigh fading and a non-fading component, giving a 'Riceian' or 'Nakagami–Rice' probability distribution. A situation which needs to be considered in theory, although possibly rare in practice, is the reception of two non-fading components of slightly different frequencies or with time-varying differential delay (see Sections 4.6 and 4.7), while for the analysis of measured characteristics the 'log-normal' law is useful.

However, measurements suggest that when the analysis is carried out over relatively short periods of time (i.e. a few minutes) a good approximation to the Raleigh distribution is most common for longer ranges and Riceian for shorter ranges, so that the assumption that for most purposes the Rayleigh fading and non-fading cases may be considered as the two asymptotic extremes would seem to be justified, and is convenient and mathematically simple. As argued above for the use of white Gaussian noise, the study of more complex cases is unlikely to contribute useful information in the present context.

There are several alternative methods of expressing the rate of variation of the signal amplitude:

(a) The 'fading rate' f_m is defined as the number of crossings per unit time of the signal envelope through its median value in a positive-going direction. The 'fade period' is the inverse of the fading rate.

(b) The 'RMS fading frequency' f_r is defined as the RMS of the Gaussian power spectrum (or auto-correlation function) of the fading signal. It can be shown that $f_m = 1.47 f_r$ [3].

(c) The 'fading bandwidth' or 'frequency spread' is twice the RMS fading frequency or 1.356 times the fading rate. This is also referred to as the 'Doppler spread' but this practice is deprecated as leading to confusion with the 'Doppler shift' discussed in Section 4.7.

The similarity of a Rayleigh fading signal to narrowband noise suggests a model for analysis based on Fig. 4.1 in which the fading signal is regarded as being generated by 'shot noise' (i.e. the summation of an extremely large number

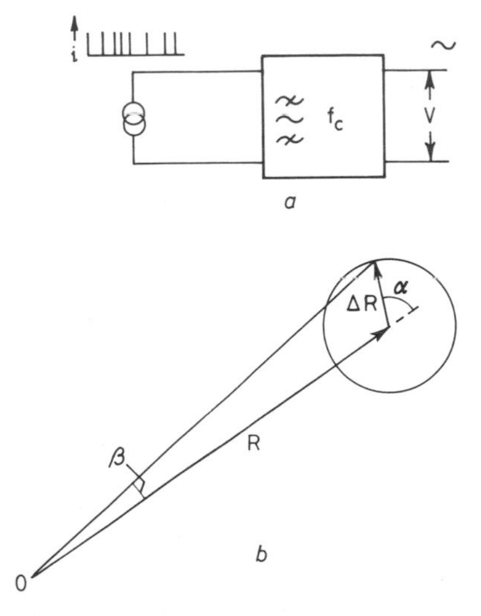

Fig. 4.1 *'Narrowband noise' model for analysis of Rayleigh fading*
a Circuit model. Gaussian filter with RMS bandwidth (single sided) $= f_r = f_m/1.47$ (see Section 4.4)
b Vector analysis of single impulse

of unit impulses arriving at random times), which is then filtered by a bandpass filter of the correct shape. Considering practical parameters of, say, a 10 MHz signal fading at one fade per second it is evident that this hypothetical filter is extremely narrow, so that the component of interest in each impulse is the sinusoidal component at the centre frequency of the filter. The phase between this component and the filter output at the instant of arrival of the impulse is randomly distributed over 2π. Thus each unit impulse contributes a sine wave component of unit amplitude and at a random phase. Then from the vector diagram of Fig. 4.1*b* it can be seen that:

$$\frac{\Delta R}{R} = \frac{\sin \beta}{\sin (\alpha - \beta)}$$

where R is the amplitude of the envelope of the output.

$$\tan \beta = \frac{\sin \alpha}{R/\Delta R - \cos\alpha} \simeq \frac{\Delta R \sin \alpha}{R}$$

The angle α is uniformly distributed, and therefore the mean modulus of $\tan \beta$ is given by

$$\left|\overline{\tan \beta}\right| = \frac{\Delta R}{\pi R} \int_0^{\pi} \sin \alpha \quad d\alpha$$

$$= \frac{2\ \Delta R}{\pi R}$$

For $\Delta R \ll R$, $\tan \beta \to \beta$, and if it is assumed that the mean rate of arrival of impulses over a time τ (short compared with the fading rate but extending over several signal cycles) is a constant 'q' pulses per second, one can postulate an 'instantaneous frequency' defined by

$$f_i(t) = q\beta(t)/\tau$$

and so the mean modulus of $f_i(t)$ is:

$$\left|\overline{f_i}\right| = \frac{2q\Delta R}{\tau\pi R} \tag{4.1}$$

Eqn. 4.1 suggests that *the instantaneous frequency* (the rate of change of phase of the signal with reference to a constant sine wave at the centre frequency of the filter) *is approximately inversely proportional to the instantaneous envelope amplitude.*

This point is illustrated by a computer simulation (based on the principles discussed in Reference 4) of a Rayleigh fading signal. Figure 4.2*a* shows the path of the end of the vector *R* of a fading signal over about three fade cycles. Fig. 4.2*b* shows the same sample, plotted in terms of absolute amplitude against time, while Fig. 4.2*c* shows the instantaneous frequency as defined above. The dotted line shows the calculated long-term median of the signal. Comparing the two curves it can be seen that when the amplitude is greater than the median the instantaneous frequency is in general less than the fading rate but that an

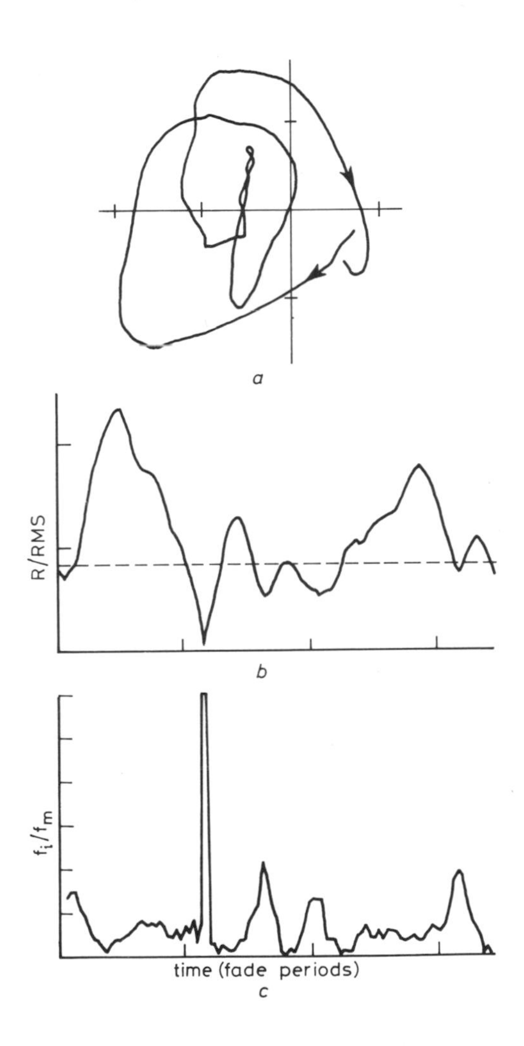

Fig. 4.2 *Computer simulation of Rayleigh fading*
a Vector
b Envelope amplitude
c Instantaneous frequency

amplitude minimum produces a short-term increase. Numerical analysis of this simulation suggests the empirical relationship:

$$\frac{R(t)f_i(t)}{\bar{R}f_m} \simeq 0.41 \text{ with variance approximately } 0.1$$

where $\bar{R}$ is the long term RMS of the signal vector and f_m is the fading rate.

A more thorough analysis is given by Rice [2].

4.5 Slow fading

The fading rate for a signal received via the ionosphere is subject to wide variation and the term 'slow fading' is defined for telegraphy purposes by the assumption that the amplitude and phase of the signal are constant during the period of reception of any one signal element. It is also assumed that the fading is flat, i.e. at any given instant the received signal amplitude is independent of signal frequency within the signalling bandwidth. Under these conditions fading itself will not cause any reception errors (if the receiving system has an infinite dynamic range). However, if added noise is present this is assumed to be not subject to the fading process, so that the signal-to-noise ratio will vary over the fade cycle and therefore the probability of error will vary also. To calculate the mean probability of error (over a period that is long compared with the fading rate), the steady-state equation must be convoluted with the probability density of a Rayleigh fading signal. The mean probability of element error is then given by:

$$p_e' = \int_0^\infty p_e(v)\mathrm{p}(v)\mathrm{d}v \tag{4.2}$$

where $p_e(v)$ is the probability of element error if the signal level is v (and for an MFSK system may be obtained from Eqn. 3.15), and $\mathrm{p}(v)\mathrm{d}v$ is the probability density of a Rayleigh distribution, which can be shown to be

$$\mathrm{p}(v)\mathrm{d}v = \frac{v}{\overline{v}^2} \exp \frac{-v}{2\overline{v}^2} \tag{4.3}$$

It can therefore be shown [5] that for an MFSK system:

$$p_e' = \sum_{r=1}^{M-1} \frac{(-1)^{M-1}(M-1)!}{(r+1)!\,(M-r-1)!} \quad \frac{r+1}{1+r\,(1+\bar{N}^2/T)} \tag{4.4}$$

This equation, combined as before with eqn. 3.16 to give byte errors, can be used to plot curves, as shown in Fig. 4.3, which can be directly compared with those for the non-fading case shown in Fig. 3.2. It can be seen that as would be expected the difference in performance between any two systems (in dB of signal-to-noise ratio), is very similar for the two cases but the variation of error rate with noise is far less rapid than when the signal is non-fading. Typically a 3 dB increase

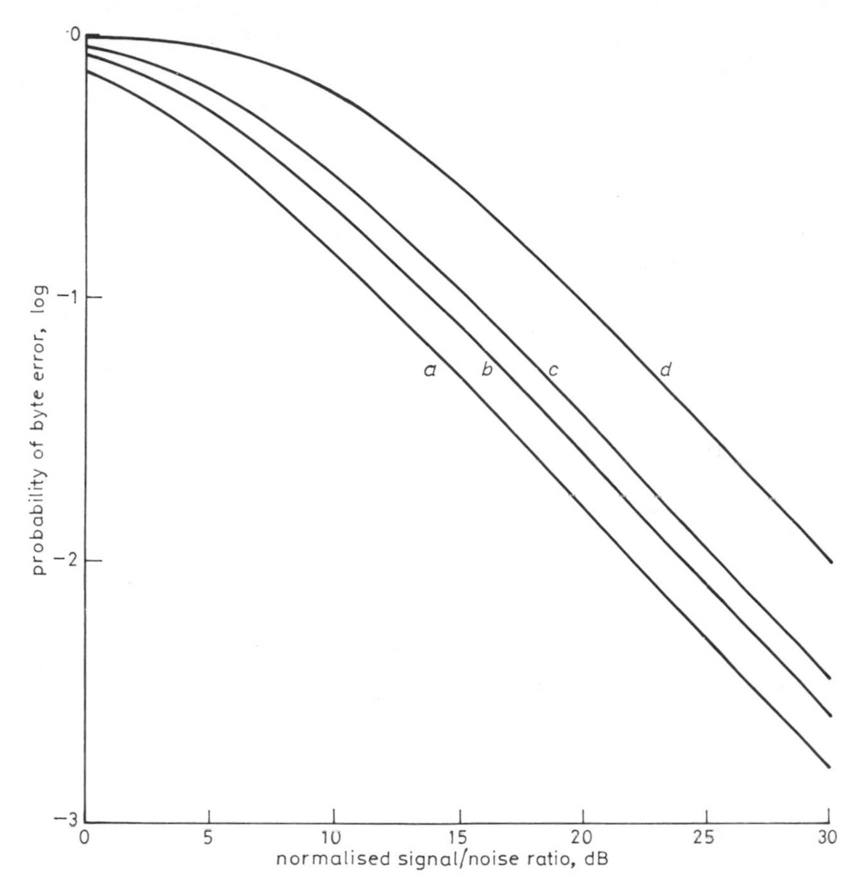

Fig. 4.3 *Byte error rates for four alternative MFSK systems*
(For parameters see Fig. 2.6)
Slow-fading signal; non-diversity reception

in noise in the non-fading case will produce an increase of 30–50:1 in the error rate whereas in the fading case the same change will produce an increase of about 2:1.

4.6 Multipath propagation

The explanation of fading given in Section 4.4 implies a 'multipath' received signal in the literal sense, but in normal practice the term is reserved for a larger scale effect whereby a receiving aerial receives two or more signals (any or all of which may be fading) with an appreciable time delay between them. The term 'appreciable' may be defined by observing that, as a rough rule-of-thumb, multipath effects are negligible unless the maximum differential delay between the signals (referred to as the path time delay, PTD) exceeds a few per cent of the length of the signalling element, or a few per cent of the inverse of the signalling bandwidth. These limits reflect the fact that multipath propagation has two distinct and independent effects on the signal, each of which must be considered.

4.6.1 Timing error

If two components are received separated in time by a delay, then a clock waveform at the element repetition rate and adjusted to be in a fixed phase relationship with one component will be out of adjustment with the other by this amount. If either signal is fading so as to be at times above or below the level of the other the effective timing of the local clock will be seen to fluctuate with reference to the received signal transitions. An ideal synchronising system would follow these fluctuations instantaneously and if this cannot be achieved then the system must allow for some degree of timing error.

Most authorities quote 'typical' figures for the PTD on h.f. links as being 2–3 ms with delays greater than 4 ms considerably less likely, but these values should be treated with caution as being applicable mainly to the long-distance links which dominated early h.f. research. In fact, the observed values of PTD rise rapidly as the range is reduced and for shorter ranges (50–400 miles) values of 10 ms have been reported [6] and over 25 ms observed in the author's organisation (quite apart from the special case discussed in Section 12.2).

4.6.2 Selective fading

Consider a received signal consisting of the sum of two components with equal amplitudes and a differential time delay τ. Then:

$$\begin{aligned} v &= A \sin \omega t + A \sin \omega (t+\tau) \\ &= 2A \cos (\omega\tau/2) \sin \omega (t+\tau/2) \end{aligned}$$

Fig. 4.4 shows the peak amplitude of v as a function of ω. If, for instance, the path time delay equals 1 ms and the radio frequency is such that exact cancellation of the two signals takes place, then a shift of frequency of 500 Hz in either direction will give a maximum signal and a shift of 1 kHz will give a null again. This is the so-called 'selective fading' effect. Even in non-ideal conditions, the result of a 1 ms PTD is that the fading pattern on two frequencies spaced by 500 Hz is largely uncorrelated or shows negative correlation.

The condition is of course not stable or static. At a radio frequency of 10 MHz a change of path time delay of microseconds can vary the amplitude of a fixed frequency signal through several nulls. Again, statistics are scarce for h.f. but as a guide, the CCIR report on the design of a channel simulator [7] suggests that it should be capable of producing a 'spectral width of a single selective fade' (presumably the frequency interval between nulls in Fig. 4.4) of 0.1 to 1.2 kHz and that these should be capable of being moved through the spectrum at 0.5 to 2 kHz/s. Other published work suggests that rates of movement of nulls of about 0.1 to 0.3 kHz/s are not untypical.

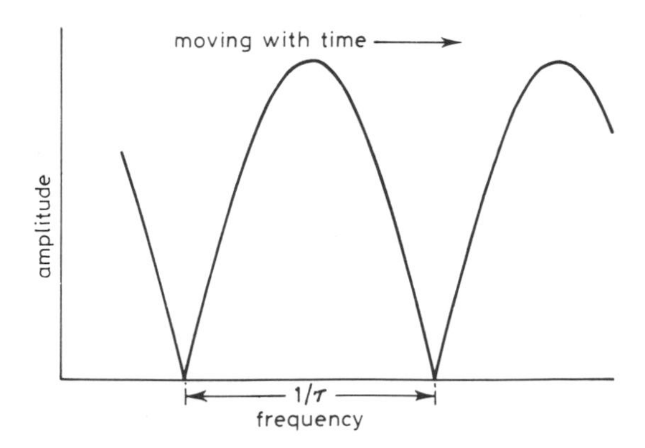

Fig. 4.4 *Selective fading effect of multipath reception*

While such effects can seriously distort wide band analogue signals such as radio telephony, the main effect on an FSK or MFSK system is to add a frequency conscious component to the fading pattern so that as the signal is keyed from one frequency to another it suffers changes of amplitude at each transition (in addition to any 'flat fading' variation). If the frequency change at a transition approaches half the interval between nulls, this amplitude change may be large. This leads to consideration of 'in-band frequency diversity' techniques discussed in Section 5.2.2, in which two signals are transmitted, separated in frequency by $(I/2\tau)$ Hz, so that there is strong negative correlation between the two fading patterns, and to the 'frequency exchange' or 'filter assessor' demodulation system for FSK [8] discussed in Section 7.5.

4.7 Doppler shift

If the ionospheric layer from which the signal is being reflected is moving bodily upwards or downwards, this (and similar but more complex effects) may cause an increase or decrease in the frequency of each component of the signal as observed at the receiving station, the effect being apparent as a shift in the whole frequency spectrum of the signal, with (in principle) no other distortion. The mathematical analysis of such an effect follows that for frequency error in Section 3.5. In fact the worst of such effects are usually associated with other disturbances, such as the transequatorial flutter fading discussed in Section 4.8. Bradley *et al.* [9] implicitly associate Doppler shifts of up to about 30 Hz with flutter fading at about 20 Hz fading rate. Shifts of similar order have also been observed in the author's organisation immediately preceding a radio blackout (sudden ionospheric disturbance – the 'Dellinger effect'), confirming other work [10,11] which associates short-term shifts of 5 to 8 Hz with solar flares and other disturbed conditions.

Longer term frequency shifts under more normal propagation conditions seem to be limited to no more than 2 or 3 Hz. In the absence of better statistics, the general impression is that a system which will cope with a steady frequency shift of about 4–6 Hz will be adequate for most cases and that appreciably larger shifts tend to occur for short periods and will almost certainly be associated with phenomena which will limit signalling accuracy for other reasons.

4.8 Fast fading

The definition of slow fading used above was that the amplitude and phase of the signal are assumed constant during each element, so that for each element taken individually the relationships derived for the non-fading case are valid. If measurements of error rate made on a practical system subject to a fading signal are compared with the theoretical curves, the results normally take the form of Fig. 6.2. For poor signal-to-noise ratios giving a high error rate the agreement between theory and measurement is good, but as the theoretical error rate falls the measured error rate levels out asymptotically to a finite probability even in the absence of noise. This residual error rate increases as the fading rate of the signal increases. It is reasonable therefore to define two regions.

> 'Slow fading' as analysed above, in which the errors are caused entirely by the signal-to-noise ratio during each element.
>
> 'Fast fading' in which the errors are caused entirely by the amplitude and/or phase fluctuations of the signal, noise making no contribution.

In the intermediate region the error rate is approximately the sum of those due to the two effects. *The actual fading rates defined as fast or slow are therefore a function of the system under consideration and the accompanying noise level.*

The fading process modulates the signal (assumed to be initially an unmodulated sine wave) in both amplitude and phase, so as to produce a Gaussian power spectrum, with energy spectral density:

$$G(f) = \frac{1}{f_r} \frac{1}{\sqrt{2\pi}} \exp\left[\frac{-1}{2}\left(\frac{\Delta f}{f_r}\right)^2\right] \qquad (4.5)$$

where $\Delta \mathbf{f}$ is the frequency, measured from the unmodulated signal, and f_r is the RMS frequency of the fading rate.

The responses of two orthogonally-spaced matched filters to a fading signal at the frequency of one of them may be deduced in general terms by reference to the shapes of the two curves in Fig. 4.5. The energy received by the correct filter will not be appreciably affected by the fading unless $f_r T > 0.3$, when it will be reduced. However, the energy received by an incorrect filter is zero in the non-fading condition and will rise rapidly as the fading spreads the signal energy either side of the null.

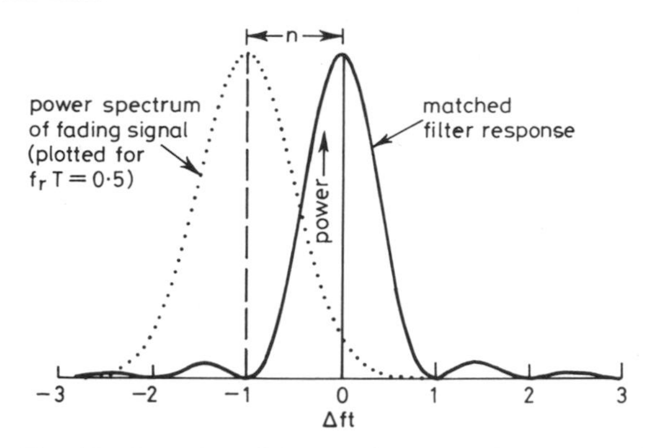

Fig. 4.5 *Derivation of theoretical error rate for fast fading*

Consider the energy W_n received by a filter at a frequency n/T Hz away from the signal. Then from eqns. 4.5 and 3.5:

$$W_n = \int_{-\infty}^{\infty} \frac{1}{fr\sqrt{2\pi}} \exp\left[\frac{-1}{2}\left(\frac{\Delta f - n/T}{fr}\right)^2\right]\left(\frac{\sin \pi\Delta fT}{\pi\Delta fT}\right)^2 d\Delta f$$

It is convenient to normalise this equation by expressing the fading rate in terms of the element rate, so let the number of signal elements per fade period be m where $m = 1/Tf_m$ so that $f_r = 1/amT$ ($a = 1.47$, Section 4.4). Then:

$$W_n = \int_{-\infty}^{\infty} \frac{amT}{\sqrt{2\pi}} \left(\frac{\sin\pi\Delta fT}{\pi\Delta fT}\right)^2 \exp\left[-0.5\,(am(\Delta fT - n))^2\right] d\Delta f \qquad (4.6)$$

Both $\sqrt{W_n}$ and $\sqrt{W_0}$ (the energy received by the correct filter) are Rayleigh fading quantities and a selection error will be produced by selecting the filter at n if a sample of W_n is greater than the corresponding sample of W_0. *If we assume that the fading of the two quantities is uncorrelated* it can be shown that the probability p_n of selecting a filter at a frequency n/T Hz from the signal (n an integer) is:

$$p_n = 1/(1 + W_o/W_n) \qquad (4.7)$$

In a complete system, and assuming there are several filters either side of the correct one, the total probability of error is then:

$$p_t = 2(p_1 + p_2 + p_3 \ldots) \simeq 2(p_1 + p_2) \qquad (4.8)$$

Thus from these equations it is possible to compute the probability of error on a fast fading signal. The curves for p_1 and p_2, and the derived approximate curves for p_t, are shown in Fig. 4.6, and also the measured values of p_t on a practical system.

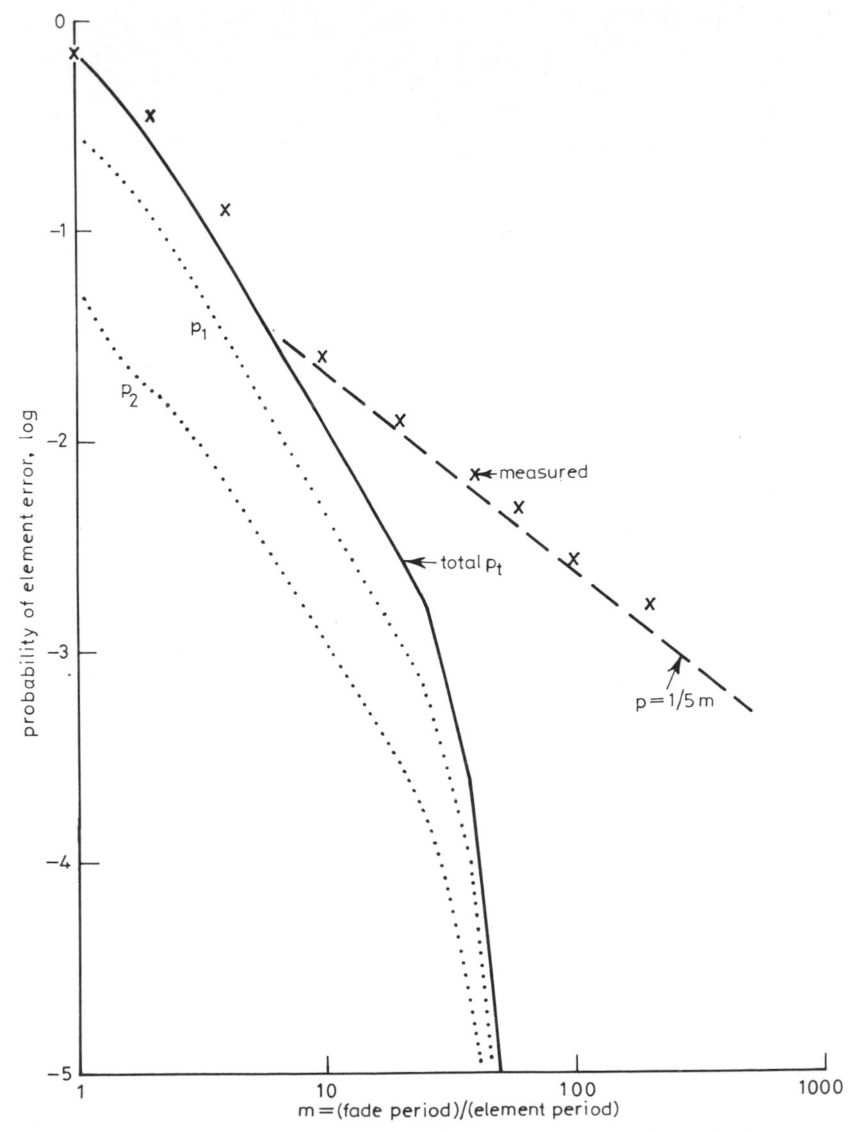

Fig. 4.6 *Element error rate of MFSK systems with fast fading signal*
Non-diversity reception

It can be seen that agreement is very close for very fast fading but at slower fading the theoretical curve tends to be optimistic. This discrepancy between theory and measurement may be explained by reference to the analysis of Section

4.4 and Fig. 4.2. When the signal amplitude passes through a minimum, the rapid variation in phase will cause a temporary broadening of the power spectrum and introduce more energy into adjacent filters. There is therefore a negative correlation between the level in the wanted filter and that in an adjacent filter (when one output is high, the other tends to be low), contravening the above assumption and causing the measured error probability to be higher than predicted by the theory. With very rapid fading the two will tend to be uncorrelated and therefore comply with the assumption.

From the measured points a convenient empirical linear approximation can be derived:

$$p_t \simeq 1/5m \tag{4.9}$$

That is, on a non-diversity MFSK system, *an element error will occur roughly once every 5 fade periods, independent of the element length.*

In principle, eqn. 3.16 should then be used to calculate the byte error rate, but assuming that the error rate is small, each element error will cause one byte error so it is also true to say that with non-diversity reception one byte error will occur roughly once every five fade periods, irrespective of the number of elements in a byte. It is also notable that $p_1 \simeq 3p_2$ so that if an MFSK system is designed with double tone spacing (i.e. the frequency interval between adjacent filters equal to $2/T$) the error rate on fast fading will be reduced to about 25% of that of the normal system. This point is discussed more fully in Section 12.1.1.

Evidently the effect of the rapid changes in phase on the responses of the matched filters is a determining factor in the errors due to fast fading. An indication of this effect can be obtained by calculations based on eqn. 3.5 of the responses of a series of matched filters when the signal has been subject to an instantaneous phase reversal during a signalling element (without any amplitude variation). It can be shown that should the reversal occur during the middle half of the element (i.e. between 25 and 75% of the element period) the response of an adjacent filter will be higher than that of the signal filter and an error will be caused. The fact that fast fading errors are created by the phase perturbations at minima in the signal fading pattern is the basis of a suggestion in Section 12.1.2 for counteracting fast fading effects.

An important corollary to the association of fast fading errors with deep fading minima is that it is immaterial whether these are caused by the mutual interference between signal components with time differences of the order of a microsecond (as in single path reception), two or three milliseconds (as in multipath reception) or at slightly differing frequencies (as could occur under Doppler shift conditions). The governing factor is the relationship between the depth of minima and the rate at which they occur. A second corollary is that under such conditions the absolute level of the received signal strength is irrelevant and an increase in transmitter power (often considered a panacea for all h.f. problems) will have no beneficial effect. Indeed, by increasing the number of viable signal paths and increasing intermodulation effects in the radio receiver it may well cause a deterioration in signalling accuracy.

When applying the above analysis to a practical design problem one should ideally have available statistics on the probability distribution of different rates of fade on the communication channel under consideration, but such statistics are not available for h.f. propagation. CCIR reports [3] suggest typical fading rates of 6 to 16 fades per minute, while other work [12] mentions fading periods between 0.5 and 60 seconds.

However, selective fading owing to multipath reception can produce faster fading and if some received signal components have suffered Doppler shift while others have not, the result will be a fade at the difference frequency. Either of these effects could contribute towards the 'flutter fading' (abnormally fast fading) which occurs frequently on some transequatorial paths. Published work [9, 13] quotes fading rates of more than 10 fades per second and observations in the author's organisation suggests that considerably higher rates (30–50 fades per second) may often be encountered on some routes. The suggestion concerning Doppler shift is advanced because there is some evidence that under such conditions the fading spectrum of the received signal is very assymetrical (see Section 12.2). It is notable that flutter fading of this nature is the only natural ionospheric condition known to the author in which the signal may be clearly audible (sometimes abnormally strong) and yet be so distorted as to severely limit the accuracy of an MFSK link.

4.9 References

1 SCHWARZ, M., BENNETT, W. R. and STEIN, S.: *Communication Systems and Techniques* (McGraw-Hill, **4**, 1966)

2 RICE, S. O.: 'Statistical properties of a sine wave plus random noise', *Bell Syst. Tech. Journal,* Jan 1948, **27**, pp. 109–157.

3 CCIR: 'Fading of radio signals received via the ionosphere', New Delhi, 1970, 2, Rep. 266–2

4 RALPHS, J. D., and SLADEN, F. M. E.: 'An h.f. channel simulator using a new Rayleigh fading method', *The Radio and Electronic Engineer*, Dec 1976, **46**, No. 12, pp. 579–587

5 SCHEMEL, R. E.: 'An assessment of Piccolo, a 32-tone telegraph system', SHAPE Technical Centre Memorandum STC TM-337, 1972, File Ref. 9980, DRIC No. P186242

6 GODDARD, F. E.: 'Description of a long-element two-tone radio telegraphy system', Signals Command Laboratories (RAF) 1962/63, Signals Order 6232, RDS Technical Minute SCL/87

7 CCIR: 'H.F. ionospheric channel simulators', Kyoto, 1978, 3, Rep. 549–1 (Annex), pp. 47–53

8 ALLNATT, J. W., JONES, E. D. and LAW, H. B.: 'Frequency diversity in the reception of selectivity fading binary f.m. signals', *Proc. IEE*, 1957, **104 B**, p. 98ff

9 BRADLEY, P. A., ECCLES, D., KING, J. W.: 'Some effects of the equatorial ionosphere on terrestial h.f. radio communications', *Telecommunications Journal*, 1972, **39**, No. 12, pp. 117–724

10 FENWICK, R. C., and VILLARD, O. G.: 'Continuous recordings of the frequency variation of WWV . . .', *Journal of Geophysical Research*, October 1960, **65**, No. 10

11 DAVIES, K., and BAKER, D. M.: 'On frequency variations of ionospheric propagation h.f. radio signals', *Radio Science*, May 1966, **I** (new series), No. 5, pp. 545–556

12 GRISDALE, G. L., MORRIS, J. G., and PALMER, D. S.: 'Fading of long-distance radio signals and a comparison of space- and polarisation-diversity reception in the 6–18 Mc/s range', *Proc. IEE* 1957, **104 B**.

13 KOSTER J. R.: 'Some measurements on the sunset fading effect', *Journal of Geophysical Research*, May 1963, **68**, No. 9

5

Diversity and Allied Techniques

5.1 Principles of diversity reception

The discussion in the previous chapter of slow fading and fast fading errors has established that both of these major mechanisms of error generation are associated with the minima of the fading envelope and therefore the behaviour of the receiving system under these conditions is of paramount importance. In general, there are three fundamentally different methods of approaching this problem:

(a) By diversity reception, i.e. the provision of two or more versions of the signal, each having a fading pattern uncorrelated with the others. This reduces both the depth and the recurrence rate of the minima on which the system must operate.

(b) By reducing the range of amplitude variation from the fading, either by an automatic gain control (AGC) system or by a process of peak amplitude limitation or clipping. An extreme case of the latter is 'zero clipping' in which the analogue signal is converted to a binary waveform, the transitions of which correspond to the zeros of the signal waveform.

(c) By designing the demodulator circuits themselves to have the largest possible dynamic range and the best possible immunity to fading effects.

In considering the relative effectiveness of these techniques it is important to study both of the error producing mechanisms.

The basic concept of diversity reception is that the receiving demodulator has available to it a number of different versions of the signal, received by different paths, or on different frequencies, or at different times. The essential requirement in all cases is that the fading patterns on the different versions are unrelated, so that when one version is fading through a minimum, there is a high probability that a workable signal will still be available on one or more of the others. The theoretical analysis of diversity operation is considered in detail in many communications textbooks [1] and will not be repeated here. The analysis is usually extended to include multiple-order diversity and it is shown that the improvement obtained decreases as the number of versions is increased (typically, two signals will give about 17 dB improvement over a single signal, four a further 9 dB, eight a further 5 dB and so on – see Section 7.7.2). This law of diminishing returns, together with practical economics, often limits the order of diversity to no more than two, and in any case, conclusions drawn from the dual diversity case can be simply extended, so this condition will be primarily considered.

An illustration of the advantage of diversity operation can be seen from Fig. 5.1, which shows the cumulative probability of the envelope of fading signals (i.e. the percentage of time for which the envelope amplitude is below a given level v).

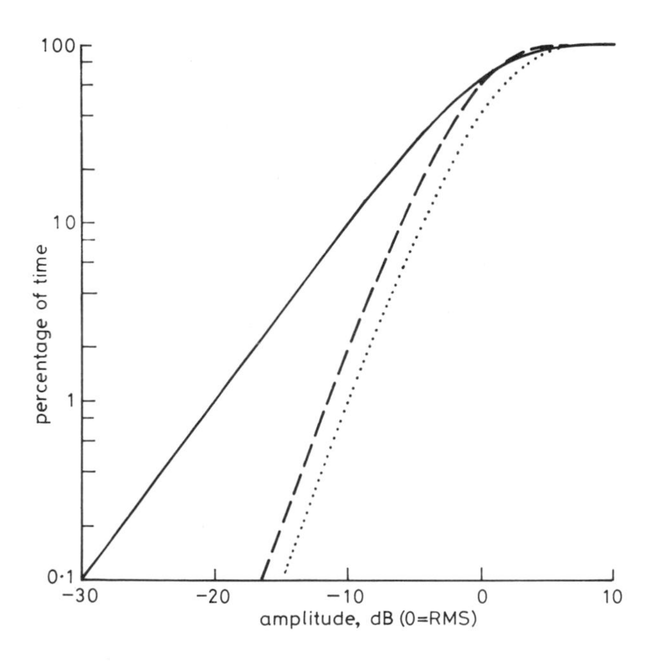

Fig. 5.1 *Cumulative probability of Rayleigh fading signals*
——— one signal
- - - - - - mean of two signals
. . . . . . larger of two signals

The relevant relationships can be shown to be:

For a single Rayleigh fading signal:
$P_1(y) = 1 - \exp(-y)$
For the instantaneous mean of two uncorrelated Rayleigh fading signals:
$P_2(y) = 0.5\ [1 - (1 - y)\exp(-y)]$
For the higher of two uncorrelated Rayleigh fading signals:
$P_3(y) = [1 - \exp(-y)]^2$

Where $y = (v/\bar{v})^2$, $\bar{v}$ being the long-term RMS voltage.

It can be seen that a signal is below −10 dB for about 10% of the time, the mean of two is below the same level for about 2% and the higher of two is below the same level for only 1%. The advantage of diversity working is therefore evident, and despite the difficulties and reservations discussed below, it is usually accepted that in any h.f. telegraphy system some form of diversity reception is essential, and that it is unrealistic to expect that the deterioration in performance evident in a non-diversity receiving system can be completely compensated for by any other techniques such as error coding, adaptive path correction and so on unless the method used implies in itself a diversity function. There are certainly practical circumstances in which it is difficult to provide two totally uncorrelated paths, but experience in the author's organisation suggests that poor diversity is still much better than none.

The analysis of diversity operation normally discusses three alternative methods of combining the signal channels:

(a) *Equal-gain combining.* The detected outputs are linearly added before comparative signal assessment or decision process.

(b) *Optimal selection.* The system selects one of the channels according to some measurable criterion, and operates only on the selected channel.

(c) *Maximal ratio (optimal ratio) combining.* The detected outputs of the channels are differentially weighted according to some measurement of the signal-to-noise ratio in each channel and the weighted outputs are then added. The weighting law is normally such as to give theoretically the highest achievable signal-to-noise ratio in the combined output.

The first two methods correspond to the 'mean' and 'higher' curves, respectively, in Fig. 5.1. The analysis of slow-fading signals in noise establishes that method (c) is ideal in all cases, method (a) gives the same performance at times when the two channels are roughly equal in signal-to-noise ratio, and method (b) is equal to the ideal when the two channels have grossly different signal-to-noise ratios. For dual diversity reception over the full dynamic range of Rayleigh fading signals the advantage of maximal ratio over either of the others is less than 2 dB.

Unfortunately, some of the conclusions based on such analysis may be difficult to justify in practice because many of the initial assumptions are not met by normal communication practice or practical conditions. These assumptions often include one or more of the following:

interference is negligible;
the errors are all caused by signal-to-noise ratio only (i.e. there are no fast fading or multipath errors);
the long term RMS values of the two signals are the same;
the noise levels in the two channels are the same, are constant and are uncorrelated with the signal levels,
the amplitude distribution of the signals is Rayleigh in all cases.

Even in the unlikely case when these assumptions are true of the voltages at the aerial terminals, the action of the AGC systems in the radio receivers (see Section 5.3) will modify the absolute levels and the amplitude distribution of the signals, and will introduce a negative correlation between the signal amplitude and the noise amplitude on each channel. In practice, the method of diversity combining used is often more a matter of practicability and economics than theoretical justification. Some attempts have been made in recent times to include maximal ratio combining in commercial equipment, but the resulting electronic circuits can be rather complex, and in view of the effects discussed above it is arguable whether even the small advantage predicted by theory is achieved. Either of the other two modes is relatively simple to implement. Note that while selection can be carried out before the detection process (provided that a criterion on which to base selection is available), addition must be of detected signals since the phase of the audio signals before detection is not coherent.

Selection diversity was chosen for the Piccolo Mark 6 (see Section 6.1), first, since the primary function of the second channel is to provide a working signal during a period when the first is passing through a deep minimum, and therefore it may be assumed that diversity operation is most necessary when there is a large difference of signal-to-noise ratio between the paths. A second point is that selection provides a nominally constant signal to the matched filters whereas in the case of addition, the failure of equipment in one of the signal paths will reduce the working signal by half, under the very conditions when the best possible low amplitude performance is required.

An important factor in the effectiveness of a selection diversity system is the criterion on which selection is made. Comparison of the absolute signal power in the channels is subject to limitation by noise, interference, and the effects of the radio AGC system. Various systems of assessing the signal-to-noise ratio by processing the phase jitter on the signal have been advocated but in many cases this must be integrated over an appreciable time in order to distinguish noise from modulation and the effectiveness of this principle under conditions of signal distortion (as distinct from noise) has yet to be established.

The method used (see Section 9.7) selects channels for each of the tone frequencies independently, selecting the higher of the two detected signals for each tone. Therefore, for slow fading signals of equal long-term power, and with the same noise level in each channel, an error will be caused only if both signals are below the critical level with reference to the noise. This system is simple to analyse by an extension of the approach of Section 4.5, and Schemel [2] shows

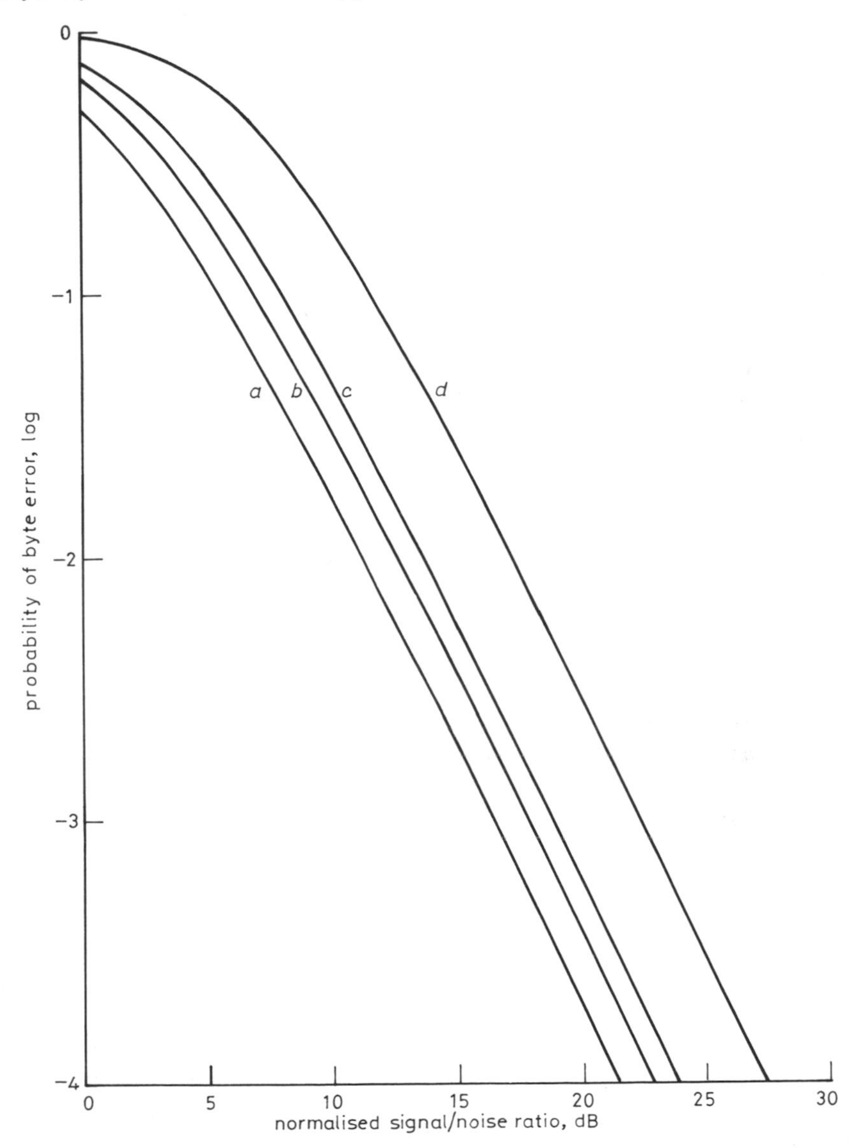

Fig. 5.2 *Byte error rates for four alternative MFSK systems*
(For parameters see Fig. 2.6)
Slow fading signal; Dual selection diversity reception

that if p_e' is the probability of element error on a non-diversity MFSK system receiving a fading signal (calculated from eqn. 4.4), the corresponding probability of error on a dual selection diversity system of this type is given by:

$$p_e'' = \frac{(2 + p_e' T/\bar{N}^2) p_e'}{2 + T/\bar{N}^2} \tag{5.1}$$

Curves corresponding to those of Fig. 4.3 are shown in Fig. 5.2. The difference in dB between the curves remains the same, but the error slope is steeper, a change of 3 dB giving a change of about 3:1 in error rate.

As indicated above, the effects of the different methods of diversity combining on the phase fluctuations caused by fading should ideally be considered. Unfortunately, there seems to have been very little analytical work published on this subject, and it is suggested that a detailed study may be rewarding. The analysis below is limited to the selection diversity method described above applied to an MFSK system, but the approach may be adapted to other cases.

Under fast fading conditions, the effect of phase perturbations is to reduce the response of the 'wanted' detector while increasing the responses of the others (see Section 4.8). Assuming that the effects only take place in one channel at a time an error is caused only if the distortion is so bad that the response of one of the unwanted detectors is higher than the response of the wanted signal detector *in the unaffected channel.*

It is then possible to derive an equivalent equation to eqn. 4.7 for such a selection diversity system, when (with $S = W_0/W_n$)

$$p_n' = \frac{7S + 2}{2S^3 + 7S^2 + 7S + 2} \tag{5.2}$$

As can be seen in Fig. 5.3, the computed curve and measured results show similar characteristics to the single-path case, but with a steeper slope, a convenient empirical approximation being:

$$p_t' \simeq 1/2m^3 \tag{5.3}$$

Note that this relationship is no longer proportional so that the value of C (the number of elements in a byte) will affect the performance of the system. Specifically, if one is comparing the error rate in fast fading conditions of several systems with the same overall data rate (as in Fig. 2.6) one would expect to find little difference in the error rate with non-diversity reception but for the error rate to be proportional to $1/C^2$ if dual diversity reception is used. (This conclusion is confirmed by measurements shown in Fig. 7.4).

Similar curves for equal-gain combining have not been derived but the comparison would be worth studying. Optimal-ratio combining has no advantages under fast fading conditions since the signal-to-noise ratio may be quite high on both channels, but it is interesting to postulate an equivalent system in which the weighting is varied according to the mean rate of change of phase in each channel.

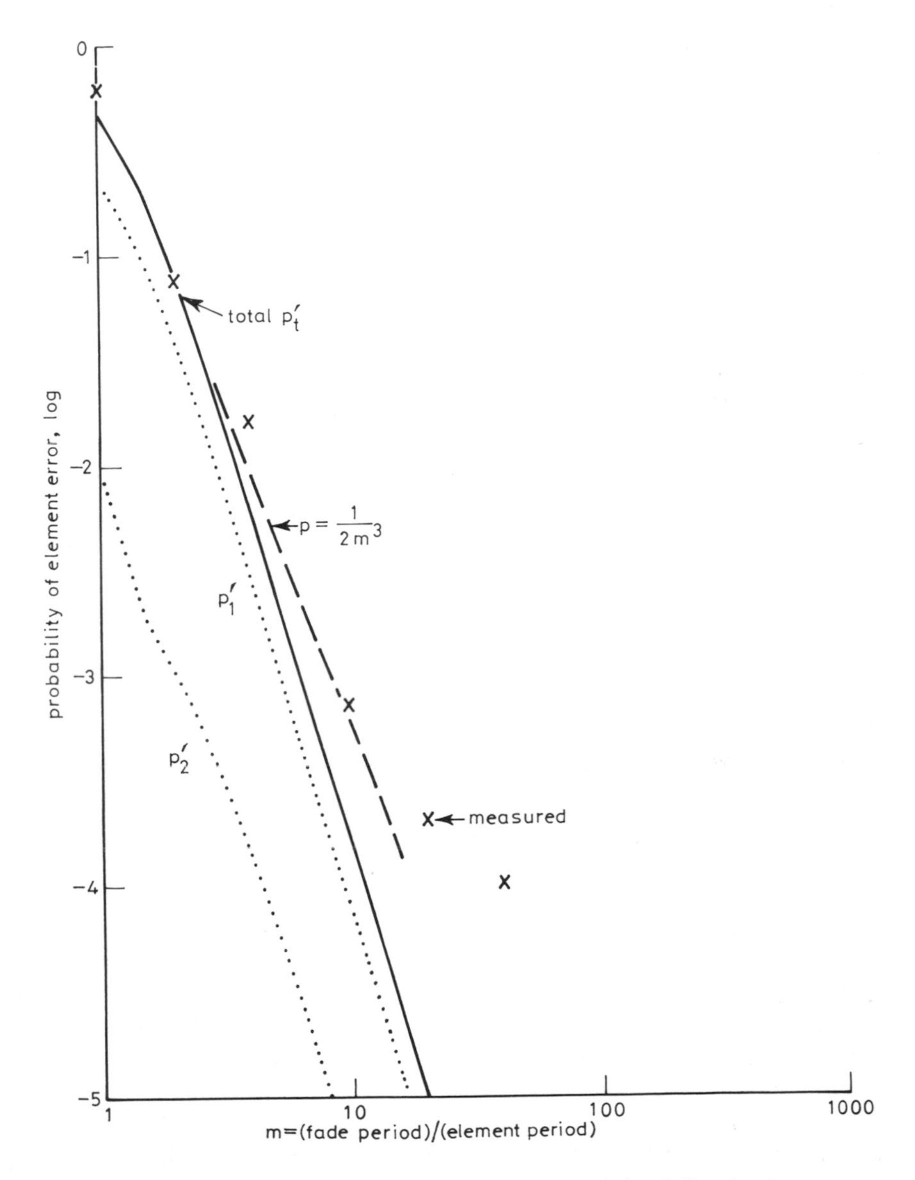

Fig. 5.3 *Element error rate of MFSK system with fast fading signal*
Dual selection diversity reception

5.2 Methods of Diversity

In considering the different methods by which independently fading signals may be obtained there are at least five different possibilities.

5.2.1 Out-of-Band Frequency Diversity

The signal is transmitted simultaneously on two widely separated radio frequencies, the separation being anything from tens of kilohertz to several megahertz. This system gives excellent diversity in that the fading on the two signals is completely non-coherent. However, it is an expensive method, requiring in most cases duplication of radio transmitters and receivers, and it doubles the problems of interference.

5.2.2 In-Band Frequency Diversity

The two signals are transmitted simultaneously within one frequency band (normally 3 kHz). Duplication of radio equipment and aerials is not necessary but the single transmitter is operated inefficiently, as described in Section 7.6.2, the receiver bandwidth must be wider than optimum, and the frequency separation may be insufficient to ensure a low correlation in the fading patterns of the two signals.

A special case where the last factor does not apply is when this type of diversity is used to combat the selective fading effects of multipath propagation (see Section 4.6.2). If the frequency separation between the two signals is equal to half the inverse of the multipath delay, then conditions causing a selective fading null on one of the signals will cause a maximum on the second signal. The filter assessor system of FSK demodulation (see Section 7.5) takes advantage of this effect, in that it is capable of demodulating either of the two frequencies independently and so has an in-built diversity action. Note, however, that since either or both of the components of the multipath signal may be independently flat fading, such compensation cannot be completely effective, and also that the technique implies a knowledge of the current multipath delay and the freedom to select the frequency separation between signals accordingly (see also Section 7.5).

5.2.3 Time Diversity

Using a single transmission path, a signal may be repeated a number of times and the different versions stored in the receiving equipment and combined by any of the diversity methods described above. Depending upon the time delay between the repetitions of the same signal, the fading patterns on the different versions can become completely non-coherent and an excellent diversity action obtained. The two major disadvantages of this technique are the considerable reduction in the data rate in the signalling channel and the increase in complexity of the receiving equipment (which must in principle store analogue voltages or waveforms). The increase in the decoding delay may also be a nuisance in some applications. However, this technique has been employed in some modern telegraphy modems in conjunction with in-band frequency diversity to give a very powerful and effective means of communication, albeit with considerable penalties in transmitter power, bandwidth, and decoding delay (see Section 7.5).

5.2.4 Space Diversity

A single transmitted signal is received on two aerials which are spaced some distance apart, each being fed to a separate radio receiver. The effectiveness of the diversity action relies on the aerials being sufficiently spaced for the fading pattern at the two aerial sites to be substantially uncorrelated, and early published work on this subject suggested that for completely non-correlated fading this spacing should be of the order of several wavelengths, preferably at right angles to the direction of propagation of the signal. Work in the author's organisation has suggested that this is a counsel of unnecessary perfection and that in fact quite effective diversity can be obtained with much smaller aerial spacings. It has been shown that an appreciable and worthwhile improvement in telegraphy accuracy can be obtained using two whip aerials spaced as little as 7 metres apart, operating on frequencies of 8 MHz and higher.

In the general discussion of diversity systems throughout this book, space diversity is assumed unless otherwise stated; for example, the term 'single aerial' is often used to imply 'non-diversity'.

5.2.5 Polarisation Diversity

The process of reflection from the ionosphere which is the basic cause of fading of an h.f. signal also rotates the plane of polarisation of the reflected wave by an arbitrary and variable amount, so that whatever the plane of polarisation of the signal transmitted from an aerial, that received at a long distance after ionospheric reflection is virtually randomly polarised. Thus the fading patterns on two aerials which are designed to receive signals with 90° difference in polarisation are almost completely uncorrelated [3]. Many of the advantages of space diversity may therefore be obtained without the physical spacing of aerials, by using two aerials crossed at right angles in the vertical plane. Since it is preferable that the two diversity signals should be approximately equal in mean signal-to-noise ratio, a configuration of two aerials in the form of an X, each polarised at 45° to the horizontal, is preferable to one with horizontal and vertical aerials. This type of array is particularly suited for short whip aerials.

5.3 Automatic gain control

The very wide range of signal levels which may be input to a radio receiver makes it essential that some form of AGC system should be incorporated in it and the characteristics of this are largely beyond the control of the modem designer. The following discussion therefore concerns only the possible use of an AGC system in the demodulator, operating on the audio signal.

Such a system may be considered as functioning in two stages. First a control voltage is generated proportional to the amplitude of the output signal and then this is applied to control a variable-gain device in the signal chain. The description in Chapter 9 of the matched filter used in MFSK detection indicates that the output of each filter is a unipolar DC voltage proportional to the amplitude of the signal component within the bandwidth of that filter. There is, therefore, no difficulty in deriving through a diode network a voltage proportional to the highest filter output (either the mean of the waveforms of Fig. 2.5 or the peak voltages at the instant of comparison). This is an excellent control function with the minimum of noise fluctuation and subject to the sole disadvantage that it must be smoothed over several elements. There is also no difficulty in applying this voltage to a variable-gain amplifier in the main audio chain. The early models of Piccolo employed such a system, and apart from the difficulty described in Section 8.3*e*, this has proved effective. Further experience and deeper study of the requirements, however, suggest that such a circuit is neither necessary nor advisable.

In the first place one should assume that there is no manual adjustment by the operator of the audio level either within the radio receiver or the MFSK modem, on the grounds that there is little point in having an AGC system if its effect can be nullified by mis-operation of an external control. This means that the radio

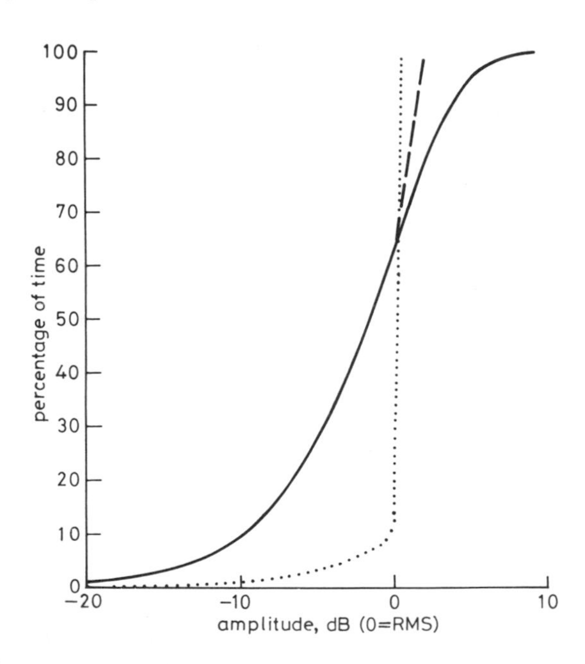

Fig. 5.4 *Cumulative probability of Rayleigh fading signal with restricted dynamic range*
——— as received
- - - - - - with peak clipping
. . . . . . with AGC

receiver is adjusted once for all to give a standard audio output on a 'typical' clean signal, and that all variations around this level are compensated for by AGC systems or accepted by the dynamic range of the equipment.

The AGC system of a modern communication receiver is extremely effective, a 100 dB increase in input signal level above a threshold (normally 0.5 to 2 microvolts) typically producing a change of 3 dB in audio output level. The dotted line in Fig. 5.4 shows the effect of such a characteristic on the cumulative probability of a Rayleigh fading signal with RMS level 10 dB above the threshold. It can be seen that the signal level is held virtually constant except for really low minima. However, since the action of the AGC is merely to alter the gain of the receiver, it will not affect the signal-to-noise ratio in a narrow band around the signal, and therefore as the input signal falls, so the signal-to-noise ratio on the output will fall correspondingly.

In view of the effectiveness of the AGC it may be assumed that any major changes in audio signal level are due to one or more of three causes:

(a) The input level of the r.f. signal at the receiver aerial input has fallen below the threshold.

(b) The fading of the signal is occurring at a rate which is too fast for the AGC to follow.

(c) Noise or interference within the i.f. bandwidth (in which the radio receiver AGC voltage is derived) has increased the AGC control voltage and so depressed the gain of the receiver and reduced the output of the wanted signal. This limitation will operate whenever the signal fades appreciably below the noise level, the AGC losing full control and in effect trying to maintain constant the total signal-plus-noise.

The effects of (a) and (c) will always be to reduce the level of the output signal, and therefore it may be assumed that large increases above the level determined by the long-term RMS of the input will not occur unless the signal level is fluctuating rapidly.

It is therefore arguable that, provided that the dynamic range of the demodulator circuits is such as to continue to operate linearly and efficiently down to the level of the lowest background noise and interference to be expected, an audio AGC system is unlikely to produce an improvement in performance in noise.

Another important factor is the effect of an AGC system on the 'fast fading' error mechanism. Without such a system, the mean signal level in the immediate vicinity of a low amplitude minimum will be considerably less than the RMS level and therefore the contribution of this period to the total energy accumulated in a matched filter will be correspondingly reduced. The addition of a powerful AGC system will attempt to counteract this tendency and will increase the gain near to or immediately after a null and so increase the effect of the phase perturbation on the matched filter output. In the limit, with an ideal system, the situation would approach the experimental conditions described in Section 4.6, in that the fluctuations in phase in the region of the null would not be affected, but the amplitude would remain virtually constant. It would seem intuitively possible that an AGC system may therefore increase the probability of errors due to fast fading. No experimental or analytical work has been carried out to justify this assumption, and again it would seem to deserve further analysis.

Summarising, the inclusion of AGC in the radio receiver is essential in order to control the gross changes in signal level at the aerial input and to reduce intermodulation problems, but once the mean level of the output signal has been stabilised in this way, there would seem to be no case for further control in order to reduce the dynamic range of the output signal due to shorter-term variations. Such arguments against the use of AGC in the demodulator system seemed sufficient to justify its omission from the Piccolo Mark 6, the design concentrating on the extension of the dynamic range of the matched filter system to as low an amplitude as practicable.

5.4 Peak limiting and zero clipping

Where an audio signal consists of a single tone frequency of variable amplitude, it is possible in principle to use the process of zero clipping to limit this variation. It is sometimes assumed that this will establish a constant level of audio and improve the signal-to-noise ratio. Neither of these beliefs is completely true.

The effect of an ideal zero clipper is to produce a binary signal, the transitions of which correspond in time to the zero crossings of the input waveform. If the audio signal is an undistorted sine wave with no added noise, then for any input level the output is a 50:50 square wave of constant amplitude. However, if the signal is distorted assymmetrically before clipping, the output waveform will not be 50:50 and the amplitude of the fundamental component will be reduced, and will vary with the degree of distortion. If noise is added to the signal this will cause a timing jitter on the transitions, generating side bands on the output signal and reducing the amplitude of the fundamental component. One method of envisaging these effects is to note that if the DC component is removed, the output power from a zero clipper is constant for any input waveform, and therefore the power in any spurious components generated must be subtracted from the fundamental.

Detailed analysis will show that if the input signal-to-noise ratio is high, that at the output is improved, but if it is low, the reverse takes place, and there is a relatively critical value of input signal-to-noise ratio at which the output deteriorates comparatively sharply. Following eqn. 3.15 it was suggested that the error rate of an MFSK system begins to rise sharply if the noise within the bandwidth of one filter exceeds about one-third of the signal level, i.e. if the signal-to-noise ratio in the signal filter is less than about +10 dB, so that for a system with M filters, the signal-to-noise power ratio in the total bandwidth of the audio input at breakdown is less than $10/M$. This means that for many practical values of M, the noise power in the full bandwidth can exceed the signal power and the signal will still be acceptable. (The 32 tone Mark 2 and Mark 3 Piccolos can give accurate data on a signal which an experienced Morse operator is unable to detect by ear). Therefore it is inadvisable to attempt to extend the dynamic range or to control the audio amplitude by zero clipping.

However, it is quite practicable to limit the high amplitude excursions of the signal by symmetrical peak limiting, because when the signal is higher than average, the signal-to-noise ratio is also above average and thus the limiting process will not cause deterioration. An optimum limit level seems difficult to establish analytically, but measurements on the Piccolo Mark 6 suggests that it should just begin to operate on the peaks of a sine wave at about the long-term RMS level of the signal, as indicated by the dashed curve in Fig. 5.4. This is based on a Fourier analysis of a peak-limited sine wave and shows that the output at the fundamental frequency can never exceed about +2 dB on the limiting level (the fundamental component of a 50:50 square wave).

5.5 Summary and Conclusions

The importance of dynamic range, and particularly the effects on the demodulator system of low amplitude signals, has been established. The factors of importance are:

The action of the radio receiver AGC.
The action of any peak limiting system in the audio amplifiers preceding the demodulator process.
The diversity system.
The dynamic range of the matched filters themselves.

The primary function of the radio receiver AGC is to establish the long-term RMS value of the audio output from the receiver at a nominally constant level. It will also reduce the dynamic range of slow fading signals to a negligible level and therefore major increases in audio output level may be assumed to occur only for short periods. Peak limiting circuits may be included to limit any such excursion to not more than about +2 dB on the threshold level.

With ideal detector circuits (i.e. matched filters remaining equal gain and linear down to vanishingly small signals, and identical comparators with zero offset, see Section 9.4) slow fading errors may be caused whenever the signal-to-noise ratio at the aerial falls below the critical level, *irrespective of the absolute levels*. If the full gain of the radio receiver (with no AGC voltage) is sufficient to amplify thermal or atmospheric noise to such a level that the detector circuits approximate closely to this standard, there would seem no justification for an audio AGC system. This criterion is easily satisfied, and with diversity reception a dynamic range of about 20 dB below nominal RMS would seem to be adequate.

However, other factors such as fast fading, poor diversity action and signal depression through interference (within the i.f. band but not on a filter frequency) all suggest that a wider range would be advantageous provided this can be obtained without increasing cost, and the matched filter system described in Chapter 9 comfortably achieves a lower limit below −30 dB without undue design complexity. This is probably enough for there to be negligible errors due to this cause, even on a non-diversity system.

5.6 References

1 BAGHDADY, E. J. (ed): *Lectures on Communication Theory* (McGraw-Hill, 1961, Chap. 7)
2 SCHEMEL, R. E.: 'An assessment of Piccolo, a 32-tone telegraph system', SHAPE Technical Centre Memorandum STC TM-337, 1972, File Ref. 9980, DRIC No. P186242
3 GRISDALE, G. L., MORRIS, J. G., and PALMER, D. S.: 'Fading of long-distance radio signals and a comparison of space- and polarisation-diversity reception in the 6–18 Mc/s range', *Proc. IEE* 1957, **104 B**.

6

Design Procedure for an MFSK System

6.1 The FCO network

This chapter will outline a basic approach to the design of an MFSK telegraphy system for a specific application. It will take as an example the design carried out during 1978–81 in the Development Section of the Communications Engineering Department of the UK Foreign & Commonwealth Office (CED) of the Piccolo Mark 6, a telegraphy modem intended for use on the network operated by that organisation and also for commercial exploitation (for other examples see Chapter 12).

The Communication Division of the UK FCO operates a communication network between the Foreign Office in London and every British embassy in the world. Many of these links are by telephone, telex or public networks over satellite, cable etc. However, from considerations of security and reliability, some 60 embassies are operated in an h.f. radio telegraphy network engineered, installed and maintained by the CED [1]. Some of the particular restrictions under which this network operates makes it a very useful example of the difficulties of h.f. communication.

Each outstation is situated in a British embassy in a capital city (which in itself ensures a very high level of locally-generated man-made noise and interference). The building is frequently quite small and in an area where visual amenities must be preserved. The result is usually a severe restriction on the size and type of aerials which can be used. The requirement to operate occasionally from emergency supplies, and a general 'good neighbour' policy restrict the outstation transmitter power voluntarily to 500 W or less. Restrictions applied by the host country may limit the number of working frequencies available to a specific station to no more than four or five, to cover a nominal 24 hours a day service over the complete sun-spot cycle. Although most traffic is passed in routine schedules during office hours, there is an overriding requirement that high priority traffic may need to be passed at short notice and at any time of the day or night. Economics and flexibility require that the number of intermediate relay stations be reduced to a minimum, and there is normally only one operational, covering the Far East. Emergency operations at ranges over 12 000 km must therefore be feasible.

The sum total of these operational requirements leads to a service in which poor signals are the norm (a signal level of 1 microvolt at the aerial terminal of a radio receiver at the base station is not considered unduly low), and often of a standard which would be dismissed as unworkable by most commercial or military networks using orthodox systems. Despite this, the general accuracy of routine traffic during the day is high, typically better than one error in 3000 characters (without error coding).

6.2 User requirements

In many cases of the development of communication equipment, the agreement between the users and the design authority on a formal list of user requirements is a long and protracted process. The designer may find it necessary to question what he sees as unreasonable demands that the system should remain compatible with all equipment and techniques in use (no matter how obsolescent they may be), and yet incorporate highly imaginative and futuristic concepts based on the capabilities of modern technology as described by the popular media and sales literature. On the other hand, the user has a deep suspicion that the design team do not understand his problems, and wish to bend the requirements to comply with their own previous experience, preconceived notions and pet ideas. In the case under discussion, most of these problems were easily resolved since the design team formed part of the user organisation and many of its staff had served for many years in the field. It was therefore relatively simple to agree on a short list of primary requirements:

1. Communication over fully synthesised point-to-point h.f. links using dual diversity reception (either space-diversity or polarisation-diversity) with professional-standard communication receivers including special-to-user modifications where these were advisable.

2. Single channel Simplex or Duplex operation in telegraph codes ITA-2 and ITA-5 (ASCII) at a basic data rate of 10.0 characters per second in either code, so as to be directly compatible with standard 75 Bd ITA-2 and 100/110 Bd ITA-5 telegraphy equipment.

3. A target of less than one error in 1000 characters under 'normal' long-distance h.f. propagation conditions (without error coding) was assumed for

design purposes, but it was accepted that operational requirements meant that in an emergency, messages may need to be passed at much higher error rates up to the limit of intelligibility of the best of several transmissions.

4. To be capable of being operated by extended or remote control with a range of data terminal equipment, including orthodox mechanical teleprinters, electronic data stores and message switches, with or without cryptographic protection.

5. Other requirements such as error coding are considered in subsequent chapters.

6.3 Communication channel limitations

No specification was laid down by the user as regards the communication channel, other than those known to be inherent in the standard of engineering and conditions of operation of the existing h.f. network. The characteristics of the communication channel to be assumed for design purposes were therefore the responsibility of the Development Section, and, since few of the network links were less than 1500 km in length with most of them between 2500 and 8000 km, were, in general those of medium- and long-haul links. From the considerations described in Chapter 4, and knowledge of the existing network, the basic guidelines listed below were derived.

6.3.1 Bandwidth

Although, in common with most communications authorities, the channels allocated to the FCO are usually 3 kHz wide, the international pressure towards bandwidth conservancy, the greater effect of multipath propagation on wider bandwidth systems, and the advantage of a narrowband system in being able to utilise narrow gaps between interfering signals, all suggested that the bandwidth should be reduced to a minimum. Furthermore, the previous system which the unit was intended to replace had operated in a bandwidth of about 350 Hz and radio receivers with i.f. crystal filters giving this bandwidth were already in use. This was therefore taken as the maximum acceptable. The narrow bandwidth would allow operation of several independent links in one band, or frequency-division multiplexed systems if required (see Section 7.7).

6.3.2 Fading

Since short range (ground wave) working was not specifically required, it was assumed for design purposes that most signals would be fading with a Rayleigh amplitude distribution at rates of up to 30–50 fades per minute. Although it was accepted that the flutter fading phenomenon previously described would sometimes set a limit on signalling accuracy and therefore should be considered as far as possible, it was agreed that this happened on a sufficiently small number of routes and for such limited times that the performance in other respects should not be prejudiced by special variation in design to overcome this one factor.

6.3.3 Multipath Delays

Since ranges below 1000 km would be relatively rare, a maximum multipath delay of about 4 ms could be assumed for design purposes. However, the possible later extension of the network to local links from an embassy to a consulate, or operation in an emergency between one embassy and another, could not be ruled out and the ability to operate with longer time delays would be advantageous.

6.3.4 Doppler Shift

For design purposes a nominal maximum of about 4 Hz could be assumed, although a greater tolerance would be useful if it could be achieved without prejudicing other factors, particularly since this would allow relaxation of the frequency stability standards for the radio equipment (see Section 10.4).

6.3.5 Noise and Interference

No figures or specification could be envisaged for these factors. All types of noise and interfering signal would be encountered at levels which would occasionally be sufficiently high as to prevent communication.

An inherent assumption was made that the technology would be available by which the time and frequency errors in the equipment itself could be made negligible compared with those in the communication channel, specifically that the synchronising system would be accurate to limits smaller than the maximum multipath delay, and that errors in the frequencies of the audio tones and filters would be small compared with the allowable Doppler shift.

The absence of statistics on the characteristics of the communication channel meant that the normal approach of attempting to design a system to meet a given performance specification (in terms of predicted availability) was impossible. Instead, it was decided that the basic parameters should be the best compromise based on the time and frequency dispersion of the channel and the required bandwidth and data rate, while the effects of noise and interference should be minimised by careful design to push to the economic or technological limit all factors which might affect the performance in this respect. For this reason the initial concept was that the signal processing in the receive system would be predominantly analogue (bearing in mind that these conclusions were reached in 1975–77) with good linearity and the widest possible dynamic range. Similarly the fact that the receiving system would normally be working in dual diversity (and therefore less subject to very low signals) was largely ignored, first, since there would be occasions when deep troughs of fade would coincide on the two reception paths, and secondly, because in case of equipment failure, operation on a single receiver would be required.

6.4 Basic parameters and codes

Initial investigations indicated that the simpler timing cycle shown in Fig. 3.8*b* would be satisfactory, i.e. the frequency separation equal to the inverse of the element length. Such a system is defined entirely by two parameters. The element length T is the primary factor, since it must give the best compromise between the effects of time and frequency dispersion, as discussed in Chapter 3. The number of tones M is then selected to give an acceptable compromise between bandwidth, data rate, and signal-to-noise performance (there is also a cost factor - see below). Of course, the selection of the two parameters cannot be made independently, since only a limited number of combinations are acceptable, as discussed in Section 2.5.

An indication of the optimum value of T for h.f. is derived as follows. One effect of multipath reception is to generate a variable synchronising error which may in theory reach the maximum value of the path time delay in either direction. Taking the maximum value of PTD as 4 ms, and assuming that a deterioration equivalent to 1 dB in signal-to-noise ratio is allowable from this cause, then from the conclusions of Sections 3.6 and 3.7, an element length greater than 20 ms is indicated. Similarly, if a Doppler shift of 4 Hz is to give a deterioration of no more than 1 dB, then from Section 3.5 a tone interval of more than 40 Hz is indicated, suggesting an element length of about 25 ms. This would give a fast fading error rate of 10^{-3} at about 5 fades/second with dual diversity reception (Fig 5.3) and 6 fades/minute with single-aerial reception (Fig. 4.6).

From these considerations it can be seen that the 'window' in the frequency-time domain for h.f. propagation is very restricted and a rigid application of pessimistic limits would suggest that accurate communication is impossible. However, the extremes assumed occur for limited periods only, and may be confined to troublesome routes or non-optimum frequencies. Furthermore, to exceed the assumed thresholds does not cause instantaneous breakdown of communication, but only reduces the immunity to noise and interference. Such calculations should therefore be regarded as indicating an area of possible parameters, rather than imposing rigid limits. For a fixed value of M, reduction in element length will increase bandwidth and errors caused by noise.

From consideration of the curves of Figs. 3.2 and 3.5 it can be concluded that there would be little point in using a value of M less than 5 or 6, which would give the lowest normalised bandwidth. High values of M incur a large bandwidth penalty for a small improvement in performance, and practical and economic factors (particularly if analogue matched filters are used, see below and Chapter 9) suggest an upper limit of about 30–40. It is worth noting that if these limits on M and T are accepted, the optimum range of application for a single-channel MFSK system of standard design, over h.f. paths, would seem to lie in data rates between about 25 and 250 bits/s, although these limits could be extended by some of the special techniques discussed in Chapter 12. and higher data rates can of course be accomplished by frequency division multiplexing (see Section 12.4).

Let us apply these conclusions to the user requirements in this particular instance. The ITA-2 code has an alphabet of 32 characters and thus, from Section 2.5, to send 10 characters/s requires a data rate of 50 bits/s. Similarly, the ITA-5 code has an alphabet of 128 and so requires a data rate of 70 bits/s. A table similar to Table 2.1 can then be constructed and all systems not meeting the requirements eliminated. This process reduces the choice to the limited range shown in Table 6.1 and the performance curves for five* of these codes are shown in Fig. 6.1. An important difference must be noted between these curves

Table 6.1 *Practicable MFSK systems for ITA-2 and ITA-5 codes at 10 characters/second (For derivation see Table 2.1; for B and B_0 see Section 3.4)*

No	Code	Alphabet	Data Rate bits/s	C	M	E %	T ms	B Hz	B_0 Hz/bit/s
1	ITA-2	32	50	1	32	0	100	370	7.4
2	ITA-2	32	50	2	6	2.5	50	180	3.6
3	ITA-2	32	50	3	4	19.0	33.3	210	4.2
4	ITA-5	128	70	1	128	0	100	1,330	19
5	ITA-5	128	70	2	12	2.3	50	320	4.6
6	ITA-5	128	70	3	6	10.6	33.3	270	3.9

and those of Fig. 3.2. The former are plotted against the normalised signal-to-noise ratio:

$$R_0 = \frac{\text{(signal energy per bit)}}{\text{(noise power per Hz bandwidth)}} \text{ dB}$$

All such curves compare systems in fundamental terms at a constant data rate. In Fig. 6.1 the signal-to-noise ratio is expressed in practical terms:

$$R = \frac{\text{(signal power)}}{\text{(noise power per Hz bandwidth)}} \text{ dB}$$

Final selection of codes suitable for the Mark 6 Piccolo proceeded as follows. For ITA-2, code No. 3 has no advantages. Its bandwidth is comparable to code No. 2, its performance is poor and the redundancy is high. Either of the other two is possible, each with specific advantages. Code No. 1 has the best performance in noise, and its long element length makes it particularly advantageous on short-distance links where the multipath delay may be high, but on long-haul

*It was not practicable to compute the curve for Code 4 since the desk calculator available was not capable of handling a number as large as 128!, which is greater than 10^{215}, but this is of little practical consequence since the bandwidth alone makes it unacceptable for present purposes.

links with a maximum PTD of about 4 ms either code would be adequate (see Fig. 7.5). The narrow bandwidth of code 2 has a decided advantage when working in crowded bandwidths on night frequencies and could largely compensate for the 2 dB loss in signal-to-noise performance against code No. 1. Under fast-fading conditions with single-aerial reception, the two codes would be similar in performance (see eqn. 4.9), while with dual-diversity reception code 2 would

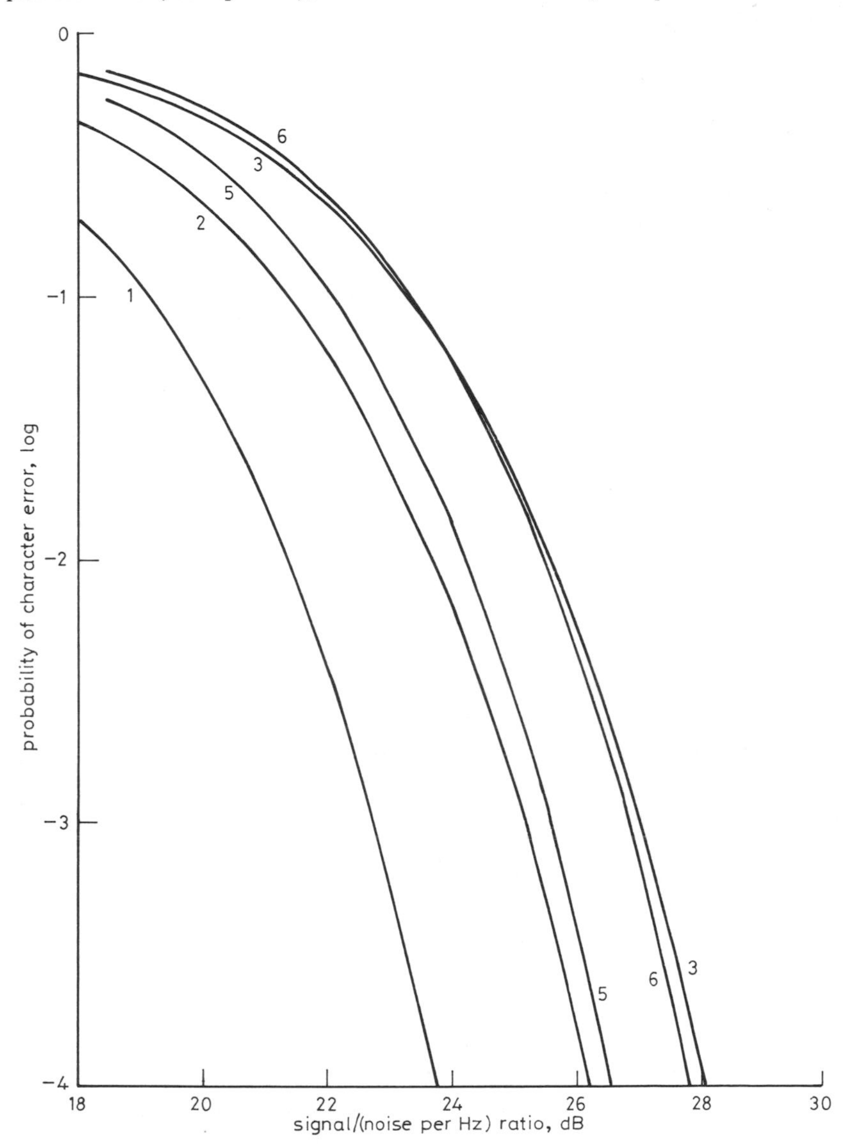

Fig. 6.1 *Character error rates for alternative MFSK systems for ITA-2 and ITA-5 codes*
(For parameters see Table 6.1)
Non-fading signal; non-diversity reception

give about a quarter of the errors of code 1 (see eqn. 5.3). Code 2 has a better resistance to Doppler shift and is less demanding on equipment frequency stability. The results of radio trials comparing the two codes are given in Section 6.5. For ITA-5, code 4 can immediately be eliminated on the grounds of bandwidth and codes 5 and 6 are equally attractive, the slight difference in performance against noise being compensated for by a reduction in bandwidth.

The early versions of Piccolo (Marks 1, 2 and 3) used code No. 1, and it has proved extremely effective. However, they were single-code equipment and suffered from a lack of flexibility and a relatively high capital cost. Study of the electronic circuit design indicated that if a two-code system is required, it is much simpler to change the number of tones than to switch between two element lengths, which also requires a change of tone interval, and so the final decision was to choose codes 2 and 5, i.e. $T=50$ ms, $M=6$ and 12. Consideration of the curves of Fig. 6.1 shows that the performance in noise of the two selected codes is almost identical.*

6.5 Measured performance

The performance of production prototype models of Piccolo Mk 6 has been measured on a channel simulator [2] and similar measurements were made by the British Post Office on a Piccolo Mk 1 [3]. The measured error rates for the non-fading and dual diversity reception of a slow fading signal are not given here since in all such tests the measured points indicate a performance within about 1 dB of the theoretical performance computed from the equations given in earlier chapters. The theoretical curve and the results of measurements for single aerial reception of a fading signal are shown in Fig. 6.2, as typical of the results obtained and also illustrating the 'fast fading' limit discussed in Section 4.8. For low signal-to-noise ratios the plotted points show a very small divergence from theory as in the other cases. At high signal-to-noise ratios, however, the measured points diverge from the theoretical, and approach asymptotically the error rate predicted from the fast fading analysis. The measured results for fast fading without noise have been discussed in Section 4.8 and are shown in Fig. 7.4 in comparison with other systems.

Measurements on the Mark 6 with a frequency offset (with noise added), were combined with the theoretical curve for errors with noise alone to produce a relationship between frequency error and increase in noise, as discussed in Section 3.5. The results show good agreement with theory, 2 Hz frequency error (10% of a tone interval) producing the equivalent of 1 dB deterioration in signal-to-noise, and 7 Hz error (35%) about $5\frac{1}{2}$ dB.

The interpretation of the measurements with multipath delay is far more difficult. To begin with, the channel simulator used produces a delay which is

*This is an interesting point, since with conventional FSK techniques an increase in data rate causes inevitably both an increase in bandwidth (because of higher modulation rate) and a deterioration of performance in noise (because of the shorter element).

absolutely constant, and therefore does not simulate the time-varying nature of the real multipath reception which is invariably (by the nature of the mechanism of multipath propagation) associated with fading. In order to introduce a degree of randomness into the delay, it is necessary to add a Rayleigh fade into either or both of the channel paths in the simulator (preferably the latter). However, if this rate of fade is too high, its effect will mask the effects of multipath for short delay times.

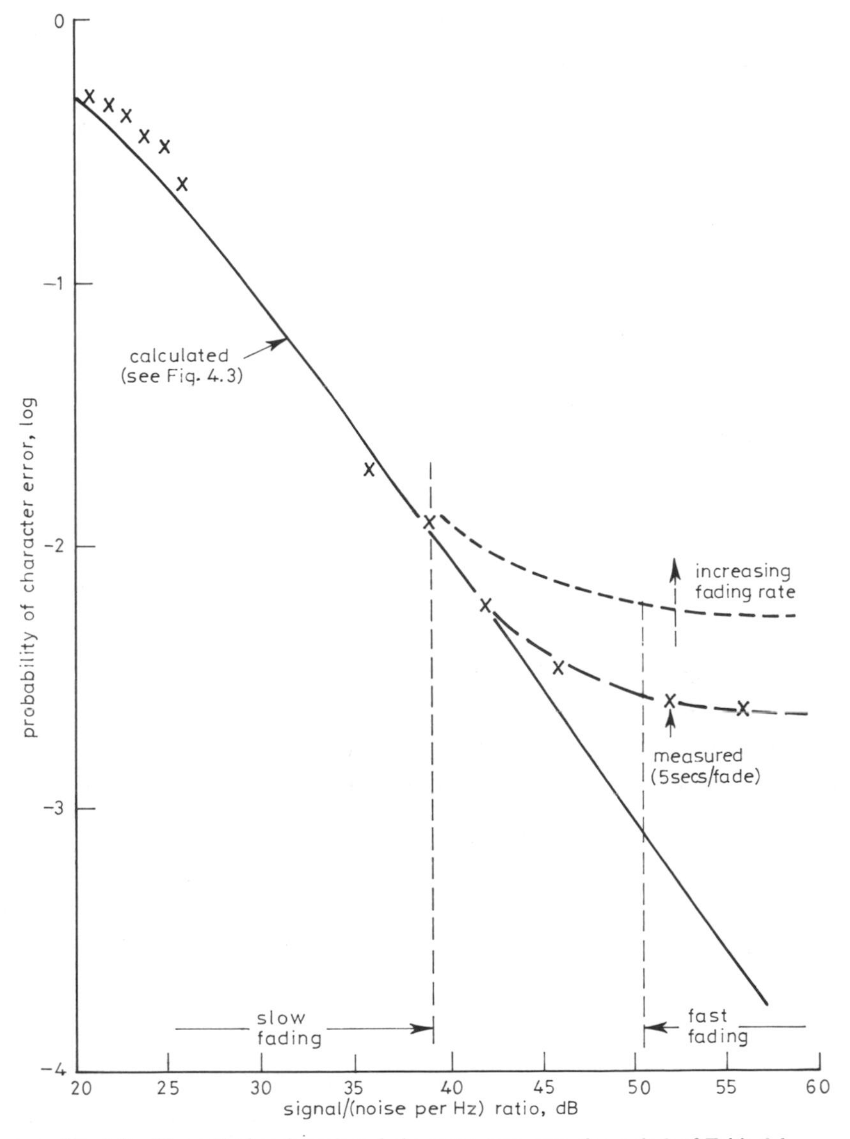

Fig. 6.2 *Theoretical and measured character error rates for code 2 of Table 6.1*
Fading signal (5 secs per fade); non-diversity reception

The second effect of multipath on the system, i.e. the production of time-varying synchronising error, is closely bound up with the design of the synchronising system rather than the matched filter detector system, and in the case of 'one-shot' synchronisation (see Section 8.5) the results will depend on the exact conditions under which synchronisation took place before the test (see Section 7.7). It is evident therefore that such measurements can only give a general indication of performance and that a direct relationship between theoretical predictions on the lines of Section 3.7 and experimental measurements is difficult to establish.

The measurements marked 'b' in Fig. 7.5 were taken on a Mark 6 Piccolo (6 tone) which was accurately synchronised by the one-shot synchronising system described in Chapter 8 on a clean steady signal, then the measurements taken with two fading paths (one delayed) of equal activity and each with the fading rate and noise indicated. The points marked 'a' were taken on an LA2028 Piccolo ($T = 100$ ms, $M = 32$) with a similar synchronising system, but in this case synchronism was carried out on the fading signal with multipath. The difference between the two curves at very low multipath delays is due to the difference in flat fading rates applied (see Figs. 4.5 and 5.3) while the difference in slope may be explained by the difference in synchronising procedures, since in the first case the multipath error can reach a maximum value equal to the PTD (and in one direction only) whereas in the second case the maximum synchronising error would depend on the conditions at the time of synchronism and would normally be not more than about half the PTD in either direction. Tentative deductions from the two curves are that the measurements tend to support the conclusions of Sections 3.6 and 3.7 and that under practical circumstances a PTD of less than 10% of an element length will produce less than 10:1 increase in error rate. This conclusion, in conjunction with the fact that the element lengths used in MFSK signalling are considerably longer than those of an equivalent binary system, suggest that MFSK techniques give a pronounced improvement in immunity to multipath effects.

As part of the process of deciding between codes 1 and 2, direct comparative trials were carried out between prototypes of the two systems on the lines suggested in Section 7.1. The tests totalled some 1900 test runs over 8 different routes, ranging from 150 km to 11 000 km. With two exceptions, the results were remarkably consistent, the character errors on code 2 being between 1.3 and 1.8 times those on code 1, and this is not inconsistent with predictions based on the analysis methods given in Chapters 3 and 4, which suggest that on a fading signal the ratio of the error rates of the two systems should be between 2:1 and 3:1 depending on the effectiveness of the diversity reception. The two exceptions were, first, that on the short range (150 km) tests the advantage of code 1 increased (to about 2.6:1), as would be expected for a signal with the observed characteristics of shallow fading, mostly at or below noise level. The other exception was the series of tests over one route which showed a reversal from the other figures, code 2 being better than code 1 by almost 2:1. This was traced to frequent interference by a signal which was on a filter frequency of the wider-band system but just outside the bandwidth of the narrower system – supporting the comments in Section 6.4.

The accuracy and consistency with which these results agreed with conclusions drawn from theoretical considerations would strongly support the contention that, while it is impossible to predict accurately the performance of an h.f. telegraphy system in absolute terms, predictions as to the relative performance of two systems (similar in data rate and bandwidth) can be confidently made. Major differences can usually be attributed to specific characteristics of the signal, so that the two principal obstacles to the design of an 'optimum' system are the lack of reliable and detailed statistics on ionospheric effects and uncertainty as to user requirements.

The economic factor in the choice of M has been mentioned,. Whereas the method of audio synthesis used to generate the transmitted signal (see Section 9.5) makes the cost independent of M, the matched filter circuit used in the receiving process (see Chapter 9) is, when constructed of discrete components or general purpose integrated circuits, appreciable in cost, thus favouring smaller values of M. The production of a special-to-purpose hybrid chip to carry out the matched filter function, or the development of fast multiplexed digital processing techniques (see Section 9.8) would considerably relax this restriction.

6.6 Character-to-Frequency Coding

Each symbol or character of the input alphabet is represented in the signal by a specific frequency or sequence of frequencies. In principle this relationship can be arbitrary and selected purely for convenience. For instance, in the Piccolos Marks 1, 2 and 3, the lowest frequency was allocated to the character A, the second to B and so on up the scale. An alternative arrangement which may be convenient in some cases would be to relate the frequency to the binary number representing the input character. When there are two or more elements to each character the operational convenience given by the simplicity of such an arrangement is largely lost and unless there is a positive factor in the users requirements which suggests a specific arrangement the choice is almost completely arbitrary, but there may be some limiting factors such as those discussed below.

Any character which, if received in error, causes abnormal or particularly damaging corruption of the received data (such as 'carriage return' on page copy) should preferably not be represented by the repetition of the same tone, since this condition can easily be simulated by c.w. interference.

The possibility of false or inaccurate synchronisation (see Section 8.8) can be considerably reduced if characters likely to be sent in long blocks of repetitions use frequencies well separated from those used for synchronising, and in cases where bandwidth economy requires that tone frequencies used for data must also be used for synchronising, it is advisable that the code be limited by not using combinations closely related to the synchronising signal. For instance, in the case of the two-tone synchronising described in Section 8.6 the three closest (i.e. either tone repeated or the same tones in the reverse order) should not be used. Then the synchronising requirements add to the effective signalling alphabet as discussed in Section 8.6.

The two codes selected for the Mark 6 are both redundant in that code 2 can generate $6^2 = 36$ characters, an excess of four over the requirements and code 5 $12^2 = 144$, an excess of 16. Thus with these two codes, the synchronising signal can be incorporated with no additional tones.

The unused tone combinations (other than data and synchronising signal) are referred to as non-valid characters* (NVC) and discussed further in Section 10.2.

Such restrictions are relatively minor and in fact the code used for the Mark 6 (developed before the availability of suitable ROM integrated circuits) is based on simplicity of conversion using binary combinatorial logic. An additional requirement in this particular case was imposed by dual-code operation. The ITA-2 (5-bit) code is treated as a subset of a 7-bit code in which elements 6 and 7 are both Mark (binary 1). The conversion algorithm is designed so that all such characters generate only the middle six frequencies (tones 3 to 8 inclusive). The synchronising signal consists of the two centre tones (tones 5 and 6), so that a change of code needs no switching of the coding, decoding or synchronising circuits, but only appropriate changes in the input and output data clocks.

6.7 References

1 BAYLEY D., and RALPHS, J. D.: 'Piccolo 32-tone telegraph system in diplomatic communication', *Proc. IEE.* 1972, **119**, (9), pp. 1229–1236

2 RALPHS, J. D., and SLADEN, F. M. E.: 'An h.f. channel simulator using a new Rayleigh fading method', *The Radio and Electronic Engineer*, Dec. 1976, **46**, No. 12, pp. 579–587.

3 WHEELER, L. K., and LAW, H. B. 'The performance of a 32-tone system subject to noise and fading'. GPO (now British Telecom Research Centre), March 1964, Report No. 20879

---

*It is an interesting possibility that this subset could be used as an auxiliary communication channel (for instance for remote control purposes) and this idea has been investigated. However, it was found that the attempt to use this additional information capacity resulted in a considerable increase in circuit complexity and in this case the benefits were more academic than practical. Note that the redundancy involved by not using these is not 15/144 or approximately 10%, as may at first appear, but, from the definition of redundancy, is (log 144/log 129) – 1 or approximately 2.3% – see eqn. 2.7.

7

Comparison with other techniques

7.1 Introduction

This chapter attempts to place MFSK techniques in perspective with other principles, both old and new. Its aim is to answer the question 'Given the problem of designing a totally new telegraphy modem to meet a given set of user requirements, what would be the optimum techniques to use?' It does *not* attempt to answer the more pragmatic problem as to which is the 'best' of existing equipments or systems. A full analysis of all possible systems and their variations is, of course, impracticable, and attention is of necessity limited to representative techniques and particularly those on which measured performance characteristics are available. The data presented includes the theoretical performance of different ideal demodulators under noise conditions. The measured performance of some practical equipment is compared, using published material and measurements made within the author's organisation, and the results are given of some comparison trials over radio circuits. Finally, conclusions drawn from the whole chapter are summarised. Consideration is mainly limited to a single channel system, some aspects of multiplexing being considered in Section 7.7.

Theoretical comparisons are limited to the cases of non-fading and slow flat fading in noise. The author is not aware of any systematic analysis of fast fading, multipath and synchronising effects on binary systems on the same lines as Section 3.5 to 3.8 and 4.8. For comparison in these fields, and an estimate as to how far the theoretical noise figures can be achieved in practice, it is necessary to have recourse to measurements made in the laboratory on actual equipments, using some form of channel simulator, or on measurements made over actual radio links. In either case it should be borne in mind that the equipment tested may not represent the ideal envisaged in the theoretical analysis. The legitimate requirements of the user may have put a strict limit on the capital or development costs, or required the ability to work with obsolescent terminal equipment, or with traditional operating and engineering methods which will not allow the full realisation of the capabilities of the demodulation techniques. The existence of a recognised standard (such as an MIL specification or CCIR recommendation) can have a particularly paralysing effect on further development, even though requirements change and technology improves. Such comparisons should therefore be treated with some care. The situation is further complicated by the fact that many modems have a number of alternative modes of operation, including provision for in-band frequency diversity, error correction coding and so on. Since in principle such additional techniques may be applied whatever basic modulation system is employed, these factors have been largely omitted in the analysis here. Similarly, in discussion of phase shift keying, only two-phase systems are considered, the results for quadrature modulation being directly deducible. The subject of error coding is discussed in detail in Chapter 11.

A channel simulator [1] is a laboratory equipment in which a test signal may be subject to artificially-generated effects which are mathematically-defined equivalents of those suffered by a signal in its passage through a real communication channel (such as fading, added noise etc.) Unfortunately, a number of techniques are available in this field, and since most such equipments have been manufactured by research laboratories for their own use, confidence in the calibration and establishment of the accuracy of simulation of second-order effects is difficult and some results must remain suspect. There are considerable differences in the range of parameters provided, and despite attempts by the CCIR [2] to standardise on the measurements to be made, these have yet to be widely accepted and it may still be difficult to compare published results from different sources. These problems apart, such techniques are a useful method of comparing the basic performance of modems under controlled conditions bearing a realistic relationship to an h.f. radio path.

Comparison of systems by measurements on actual links is even more difficult. It is almost axiomatic that measurements carried out on a single system in isolation are virtually useless in establishing the relative performance between that system and another. Factors such as terminal sites, aerial sizes and orientations, current ionospheric conditions and so on, are not available to a second investigator, while some less obvious factors, such as the number of available frequencies and the criteria for frequency changes, are often unreported. The last point may be taken as an example of the problems involved.

The choice of working frequencies on radio trials usually follows a deliberate policy, for instance:

(a) Fixed frequency and time schedules, based on prediction, experience, or operational restrictions.

(b) Deliberate choice of 'good' working frequencies and times, such decisions being guided by prediction, experience, observations of signals, study of output errors or ionospheric soundings, with the overall aim of achieving the highest possible circuit availability.

(c) Deliberate choice of 'bad' working frequencies and times, in order to study the factors which cause breakdown of communications or to compare systems under marginal conditions.

It is obvious that the policy adopted can have a major effect on results yet it is seldom defined in the trials report.

In one case known to the author, the published results of radio trials purporting to compare the performance of a telegraphy system with and without an error correcting code apparently indicated that the use of error coding produced a very considerable improvement in all cases. Only when the implications of a single phrase in the text were studied did it become apparent that a test only took place if the receiving operator judged by ear that the signal was suitable, and in over 80% of the attempts no test results were recorded!

Evidence from radio trials can only be accepted where care has been taken to ensure that both systems have been subject to essentially the same conditions and that those conditions covered a range typical of those expected in practice and preferably including some extreme or marginally usable signals. A frequent mistake is to draw firm and positive conclusions from a short trial taken under a limited set of conditions.

7.2 Derivation and presentation of data

Many textbooks on communication theory derive equations for the element error rates* of idealised binary signalling systems with added white Gaussian noise [3]. These include both coherent and non-coherent methods of detection (see Section 2.6) and although the former is not practicable over radio circuits the results are included here for completeness. The equations themselves are often quoted in books of reference [4] and are listed in Table 7.1 under three standard signalling conditions, i.e. non-fading, slow fading with single-aerial reception (non-diversity) and the same with dual (space) diversity. The equivalent equations for MFSK systems are derived in this book and given in eqns. 3.15, 4.4 and 5.1, respectively.

It is very important to note that when comparing two binary systems, it is sufficient to compare the element error rates* but in comparing a binary system with a multi-level system it is necessary to express results in terms of a common character or byte output. The reference code used here is the ITA-2 code, with 5

* Often referred to as the bit error rate (BER) but this usage can lead to ambiguity since one binary element does not always convey one bit of information. See 'Glossary of Terms and Abbreviations'.

Table 7.1 Theoretical element error rates for idealised systems

Signal conditions	Diversity order	MFSK	Binary	
			coherent	non-coherent
Non-fading	non-diversity (1)	eqn. 3.15	$0.5\ \mathrm{erfc}(aR)^{\frac{1}{2}}$	$0.5\exp(-bR)$
Slow flat Rayleigh fading	non-diversity (1)	eqn. 4.4	$0.5(1-(1+1/aR)^{\frac{1}{2}})$ $\simeq 1/4aR$	$0.5(1+bR)^{-1}$ $\simeq 1/2bR$
	dual diversity (2)	eqn. 5.1	$0.5(1-Z^{\frac{1}{2}})$ where $Z=\dfrac{(1+3/2aR)^2}{(1+1/aR)^3}$ $\simeq 3/16(aR)^2$	$0.5(1+bR)^{-2}$ $\simeq 1/2(bR)^2$
System Parameters		A: $M=32$ $C=1$, B: $M=6$ $C=2$, C: $M=2$ $C=5$,	F: (Ideal or PSK), $a=1$ G: (FSK), $a=0.5$ For 5-bit bytes $C=5$	D: (Ideal or DPSK), $b=1$ E: (FSK), $b=0.5$ For 5-bit bytes $C=5$

R = normalised signal-to-noise ratio
= (signal energy per bit)/(noise power per Hz bandwidth)
= R_o in Section 3.3

To compare equivalent synchronous systems, use eqn. 3.16 to calculate byte error rate for 5-bit bytes

bits per character, and the MFSK codes Nos. 1 and 2 in Table 6.1 (A and B in Table 7.1) will be used as examples of the technique. Code C ('binary MFSK') is included since it can be shown to be identical to an idealised FSK system (code E). Unless otherwise stated the binary systems are all assumed to be operating in a 5-element synchronous mode (assuming perfect synchronisation) and eqn. 3.16 (with $C=5$) used to convert element errors as given by Table 7.1 to character errors. For a binary system using a $7\frac{1}{2}$-unit signal, the signal-to-noise performance is worse, as discussed in Section 2.7.

It can be seen that the equations for the non-fading condition fall into two distinct categories. The error rate of each of the binary non-coherent systems (and MFSK systems to a good approximation) is represented by an exponential

Table 7.2 *Bandwidths and demodulation factors of synchronous systems with same data rate*

System	Demod factor (dB)	Normalised bandwidth B_o (Hz/bit/s) Single channel	Frequency multiplexed (approximate minimum)
A MFSK; $M=32$, $C=1$, $T=100$ ms	−6	7	–
B MFSK; $M=6$, $C=2$, $T=50$ ms	−3.5	3.5	2.5
C MFSK; $M=2$, $C=5$, $T=20$ ms	0	5	–
D DPSK 2-Phase 50 Bd	−3	4	1
D′ DPSK 4-Phase 25 Bd	−3	2	0.5
E FSK (filter assessor), 50 Bd			
500Hz deviation (opt. for 1ms PTD)	0	12	3.4†
250Hz deviation (opt. for 2ms PTD)	0	7	3.4†
125Hz deviation (opt. for 4ms PTD)	0	3.8	3.4†
FSK (limiter discriminator), 50 Bd	0	–	1.7*

* CCIR Rec 436. 15 Channels at 170Hz spacing. Deviation 85Hz, modulation rate 100 Bd
† Twice that calculated as *

curve, so that the difference in performance between any two of them is summed up in a single figure representing the change of signal-to-noise, in decibels, required to maintain a constant character error rate. Similarly, the two coherent systems are each represented by an 'error function' curve and again the difference between them may be expressed as a single figure. To express the relative performance of a coherent and non-coherent system in terms of a difference in decibels implies some degree of approximation, particularly where the shapes of the curves diverge at high error rates. Where such comparisons are made below they apply approximately to character error rates of the order of 10^{-3}. Similar comments apply to the results for fading signals, since these are derived from those for non-fading (see Section 4.5).

The term 'demodulation factor' (DF) is frequently recommended to express the difference in performance between a given system and a reference system [4]. Unfortunately, some confusion has arisen as to what system should be used for comparison. CCIR suggests comparison with 'an idealised receiver *of the same type'* (author's italics). The original paper by Law suggesting the term [5] implies the use of the coherent standard in all cases, while a companion paper [6] used as a standard for non-coherent systems a hypothetical non-coherent system having the same performance under single channel Rayleigh fading conditions as an ideal coherent receiver. In some published papers it is not made clear whether the intended reference is an 'ideal non-coherent binary detector' or an 'ideal non-coherent FSK detector'. The use of the term in an unqualified way in work involving a wide variety of detection systems could therefore lead to ambiguity and it will only be used here in conjunction with an explicit definition.†

The DFs of various idealised systems are listed in Table 7.2, with reference to systems E and C (which are nominally identical), together with approximate figures for the normalised bandwidth (see Section 3.4.4). It is important to recognise that, while the bandwidth of some systems such as MFSK and PSK may be derived in fundamental terms related to the data rate, that of an FSK system is also a function of the frequency deviation, which is not fundamentally related to data rate (and which has, in theory, no effect on the error rate – note that the equations in Table 7.1 do not involve deviation or bandwidth, explicitly or implicitly). The bandwidth figures for FSK systems in Table 7.2 are therefore based on recommended CCIR systems [7] using CCIR recommended empirical equations [8].

The error rates measured on channel simulators of experimental, commercial or military systems are taken from the references quoted. Reference 9 is a brief report only, but the measurements on which that report was based have been kindly made available to the author by the (British) Army School of Signals. The characteristics of the signalling systems on which measurements were carried out are listed in Table 7.3, and the results with multipath and fast fading are collated in Figs. 7.4 and 7.5. Note that since these measurements were carried out in different laboratories, on different simulators, and with different test conditions, the greatest care should be exercised in interpretation and only very broad conclusions should be drawn.

7.3 Coherent systems

As discussed in Section 2.6, coherent detection is only practicable over a phase-stable channel and therefore the method cannot be used over radio circuits. An 'ideal coherent binary detector' (F in Table 7.1) represents the best performance which can be obtained with any binary detector. Over a phase-stable medium it could be realised by coherent detection of two-phase (phase reversal) keying (PSK) and it is the only system considered here which is bi-orthogonal as defined in Section 2.6.

Its performance under non-fading conditions is very similar to that of system B, the 6-tone MFSK, and any MFSK system with $M > 6$ will give a better

† There seems also to be some confusion as to whether the DF is intended to express the amount that a given practical demodulator circuit falls short of its own idealised theoretical performance (in which case it is virtually obsolete, as any modern equipment should achieve a performance within 1 dB of its theory), or to express the theoretical difference between two totally different demodulation techniques, in which case the term is confusing.

performance. Under non-diversity fading conditions the relative performance of all the MFSK systems has deteriorated slightly (about 1.5 dB) so that coherent PSK is now roughly equivalent to a 10-tone MFSK system. This anomaly may be attributed to the difference between the shapes of the error function and exponential curves (note B and F in Fig. 7.1), particularly at high error rates. The advantage of coherent detection rises rapidly as the error rate increases above

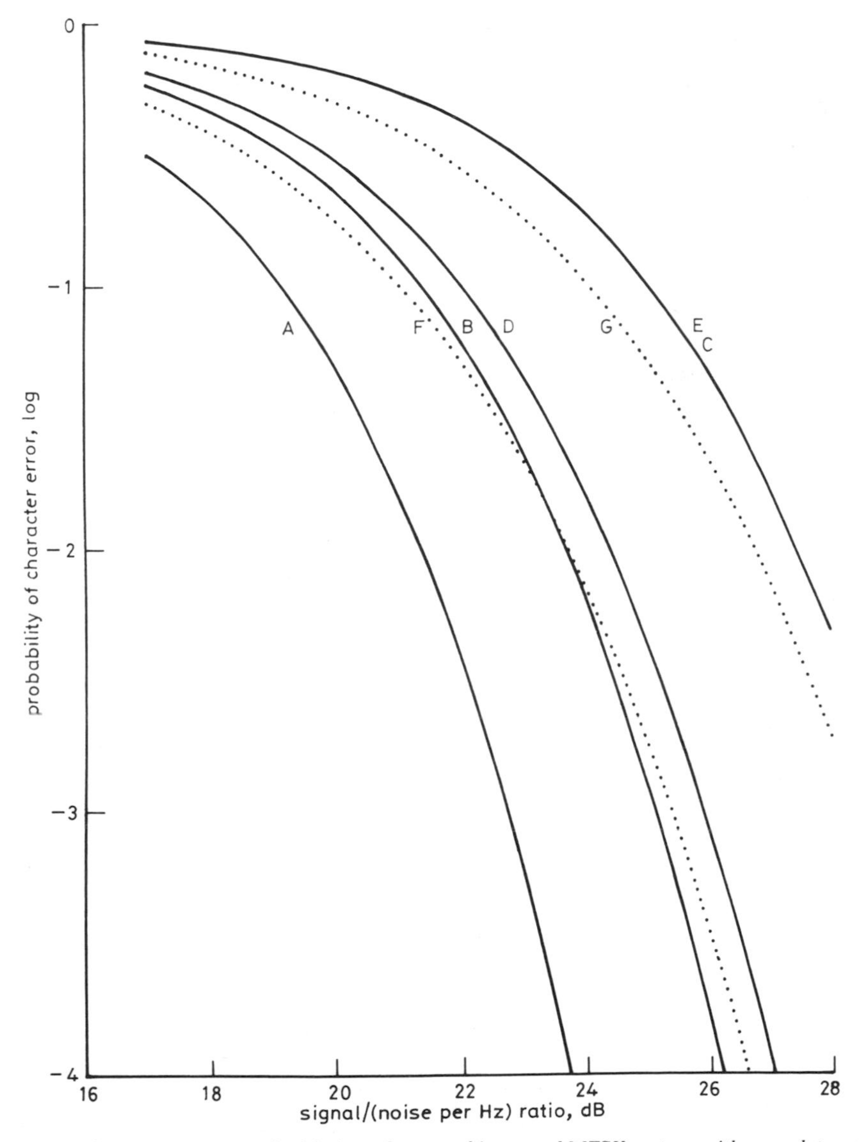

Fig. 7.1 *Character error rates for ideal synchronous binary and MFSK systems with same data rates*
(For system identification see Table 7.1)
Non-fading signal; non-diversity reception

25% (element errors above 5%). Under fading conditions most errors occur in bursts of high error density, so that improvement in this region is a controlling factor. Of course, this effect occurs in any comparisons between coherent and a non-coherent system.

There is also an apparent anomaly between the curves of Fig. 7.2 and those of Fig. 7.3, in that whereas the relative difference between any two binary systems,

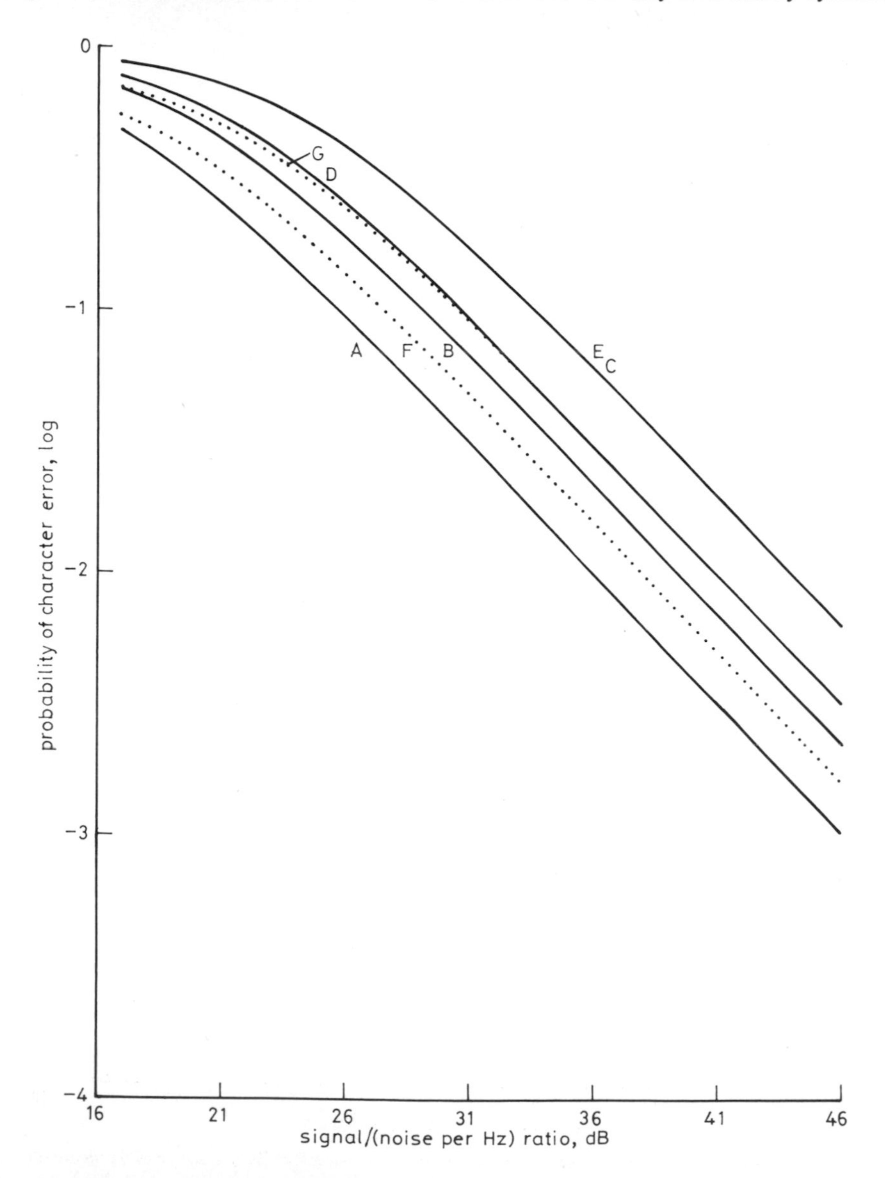

Fig. 7.2 *Character error rates for ideal synchronous binary and MFSK systems with same data rates*
Slow-fading signal; non-diversity reception

or any two MFSK systems, has remained unchanged, all three MFSK systems have apparently deteriorated about 1.2 dB. This can be traced to the differing assumptions in the mathematical analysis. The diversity performance of binary systems is usually calculated for 'maximal ratio' combination (see Section 5.1) giving the optimum performance, whereas eqn. 5.1 is based on 'selection diversity' (as used in the Piccolo equipments). The theoretical difference in performance between the two methods on a Rayleigh fading signal can be shown to be 1.5

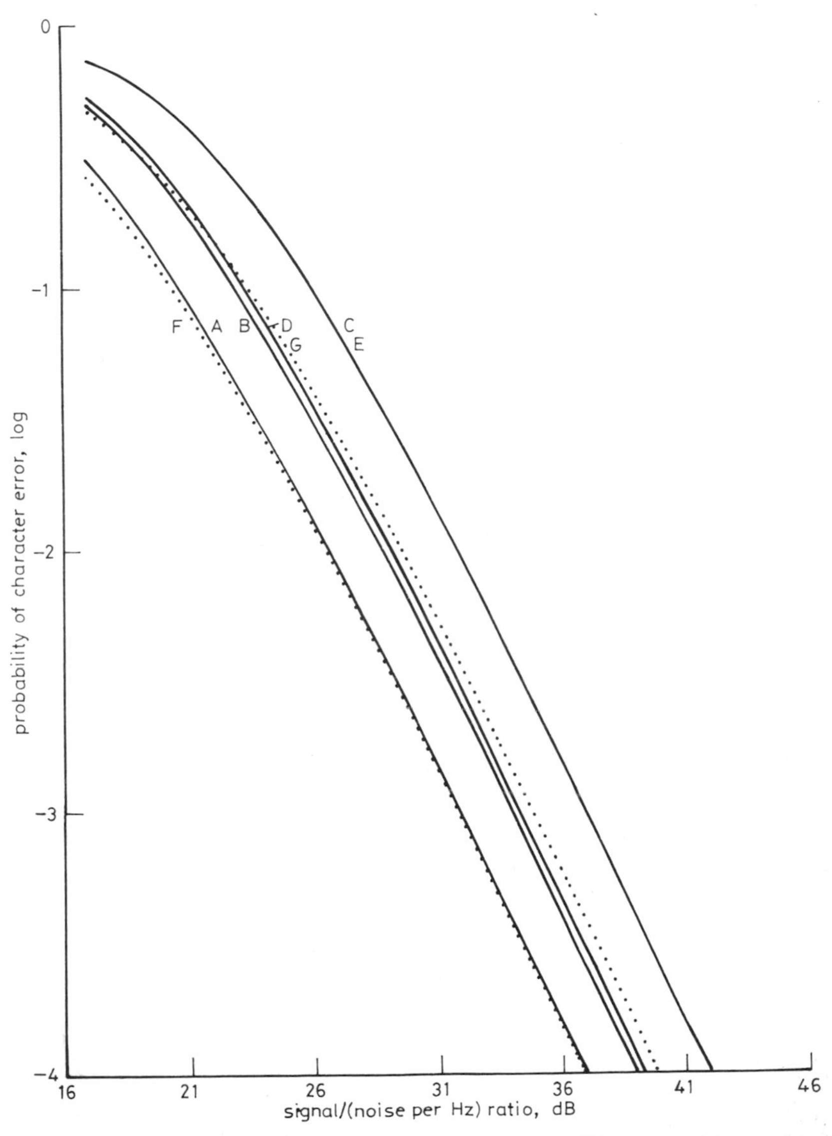

Fig. 7.3 *Character error rates for ideal synchronous binary and MFSK systems with same data rates*
Slow-fading signal; dual diversity reception

dB [10], so that it may safely be assumed that the use of the same diversity method in each case would result in the same relative performance as for single-aerial reception.

An 'ideal coherent FSK detector' (G) could be realised by coherent detection in matched filters of two orthogonally-spaced frequencies, followed by comparison of the detected outputs. Since the noise in the two filters is added, the performance is 3 dB worse than coherent PSK, and slightly worse than 6-tone MFSK under fading conditions.

Since no practical implementations of coherent systems are available, no measured results are given. A system referred to as 'coherent FSK' has been described [11], but since it operates by comparing the signal phase at the beginning and end of an element (as in DPSK, see below) the nomenclature is doubtful.

7.4 Non-coherent binary detectors and DPSK

The 'ideal non-coherent binary detector' (system D) represents the best performance which can be achieved with a binary signalling system when the detector has no *a priori* information as to the phase of the signal, but the phase is stable over any one element. This performance can be attained in theory by differential phase-shift keying (DPSK) in which a single tone is two-phase (phase reversal) modulated, the information being conveyed by the change in phase between one element and the next (e.g. binary '0' = no change; binary '1' = phase reversed). Demodulation is by coherent detection, but the phase of the reference sinusoid is determined by the received phase of the previous element. As such, the reference phase may be disturbed by noise (particularly on a fading signal) and so the performance is poorer than coherent PSK (F) by about 3 dB on fading signals, being very similar to coherent FSK and slightly worse than 6-tone MFSK.

Probably the first practical implementation of DPSK was the Kineplex equipment [12, 13], in which 20 phase-modulated tones and a reference were frequency-multiplexed into a 3.4 kHz bandwidth. This has to a great extent been superceded by a similar system [14] which uses 16 modulated tones and a reference in a nominal bandwidth of 3 kHz. The modulation parameters are similar in both cases, each data tone being modulated at 75 Bd (13.3 ms elements) in two or four phases, the latter mode giving a total data rate of $2 \times 75 \times 16 = 2400$ bits/s.

The tones are spaced at 110 Hz intervals, orthogonal for the matched filter integration period of 9.1 ms, which allows a guard time of about ± 2 ms to cater for a limited degree of multipath or synchronising error, (see Section 3.8). The provision of this guard period causes an effective power loss of 4.2/13.3 or 1.75 dB, so, as would be expected, the measured DF from the ideal DPSK is about 2.5 dB [13].

If the normalised occupied bandwidth of a single quadrature-modulated tone is calculated on the same basis as in Sections 3.4.3 and 3.4.4, the normalised bandwidth is 440/150 = 3 Hz/bit/s. However, the frequency multiplexing of a number of carriers at orthogonal intervals achieves a very pronounced spectrum economy and the normalised bandwidth of a 16 channel system on the same basis is $(19 \times 110)/(16 \times 150) = 0.87$ Hz/bit/s, probably approaching the ultimate in bandwidth efficiency (the pilot tones are ignored for this calculation).

The few measured figures on fast-fading errors (see j in Fig. 7.4) suggest that the technique breaks down very rapidly as the fading rate increases above about 0.5 fades/sec (2 secs/fade), while the data for multipath with dual diversity reception (see J in Fig. 7.5) shows, as expected, that delays less than 2 ms have little effect, but errors rise rapidly at longer delays.

The ANDEFT system [15, 16] is a frequency-differential phase shift keying system. The binary data is conveyed by the difference in phase change between a tone and an adjacent tone (40 Hz away), on the assumption that differential phase modulations due to fading will be relatively small over such a narrow frequency band. It uses a total of 66 tones spaced at 40 Hz intervals between 400 Hz and 3 kHz, 64 of which are phase modulated, the system catering for either 2-phase or 4-phase modulation. The unmodulated tones are for reference purposes. The transmitted element length is 26.67 ms (37.5 Bd) the matched filters integrating over 25 ms, leaving a guard time of about 1.7 ms, and a resulting power loss of 0.25 dB. Using 4-phase modulation each modulated tone can convey 75 bits per second and the whole system 4800 bits per second. Alternative modes are provided for 2-phase operation and for dual-in-band diversity operation.

By the same standards as previously, the normalised bandwidth of a single channel is approximately 1.33 bits per second, and again this is reduced by the multiplexed operation, although the statement in the reference that 'significant sidebands extend from 362.5 to 3018.8 Hz' (giving a normalised bandwidth of less than 0.6 Hz per bit) should be taken with some reservations as they are apparently based on the assumption that there is no significant power beyond the first null in the ($\sin x/x$) spectrum of the end tones.

The measured raw error rate for flat-fading signals in noise is quoted as being within 1 dB of the theoretical, presumably with no allowance for guard time and so is (on a normalised basis) some 1.5–2 dB better than Kineplex.

ANDEFT and Kineplex are fundamentally similar, the first requiring coherence over one element (27 ms) on two adjacent frequencies (40 Hz), and the second over two successive elements (27 ms) on a single frequency, so although no fast-fading error rates on ANDEFT are available, one would expect a very similar performance in this respect. The multipath measurements (see H in Fig. 7.5) in dual (space) diversity show that despite the longer element, the shorter guard time gives a higher error rate, although additional in-band diversity (H′) apparently gives a major improvement (but note that this data in particular must be considered with care, since the true curve will exhibit maxima and minima – see Fig. 7.10).

Doppler shift is not discussed as a separate item in this chapter mainly because, while the measured behaviour of MFSK systems follows closely on that predicted

Table 7.3 Systems plotted in Figs. 7.4 and 7.5

Ref	System	Source Ref (Chap 7)	Code	Element period (ms)	Mod'n rate (Bd)	Frequency shift or interval (Hz)	Phases
a	MFSK Pic. Mk1, Mk3, LA2028	–	32 tones 1 element ITA-2	100	–	10	–
b	MFSK Pic. Mk6 LA 1117	–	6 tones 2 elements ITA-2	50	–	20	–
c	FSK filter assessor	9	$7\frac{1}{2}$ unit ITA-2 start–stop	13.3	75	85	–
d	FSK filter assessor	9	$7\frac{1}{2}$ unit ITA-2 start–stop	13.3	75	850	–
e	FSK Filter Assessor	17	*Binary uncoded	10	100	510	–
f	FSK Limiter Discriminator	17	*Binary uncoded	10	100	80	–
g	FSK Limiter Discriminator	17	*Binary uncoded	10	100	170	–
h	DPSK ANDEFT	13 (p.318)	*Binary uncoded	26.7	37.5	–	4
j	DPSK Kineplex	17	*Binary Uncoded	13.3	75	–	4

System A = system 'a' in dual (space) diversity etc.
System H′ = system 'h' in dual (space) plus dual in-band diversity.
* Binary uncoded systems are assumed to be working 5-unit synchronous; character error rate = 5 × element error rate.

by the theory of Chapter 3, the literature yields little or no corresponding theoretical analysis or practical measurements on other systems. The frequency-correcting circuit included in many DPSK receiving equipments is apparently intended to compensate for instabilities in the radio equipment rather than the ionosphere (frequency errors of 30–50 Hz are catered for, while stabilities of 2–6 Hz are required). The behaviour of such a system may have a pronounced effect in Doppler or fast-fading situations – a subject seldom referred to in published specifications. In fact, since the occurrence of ionospheric Doppler

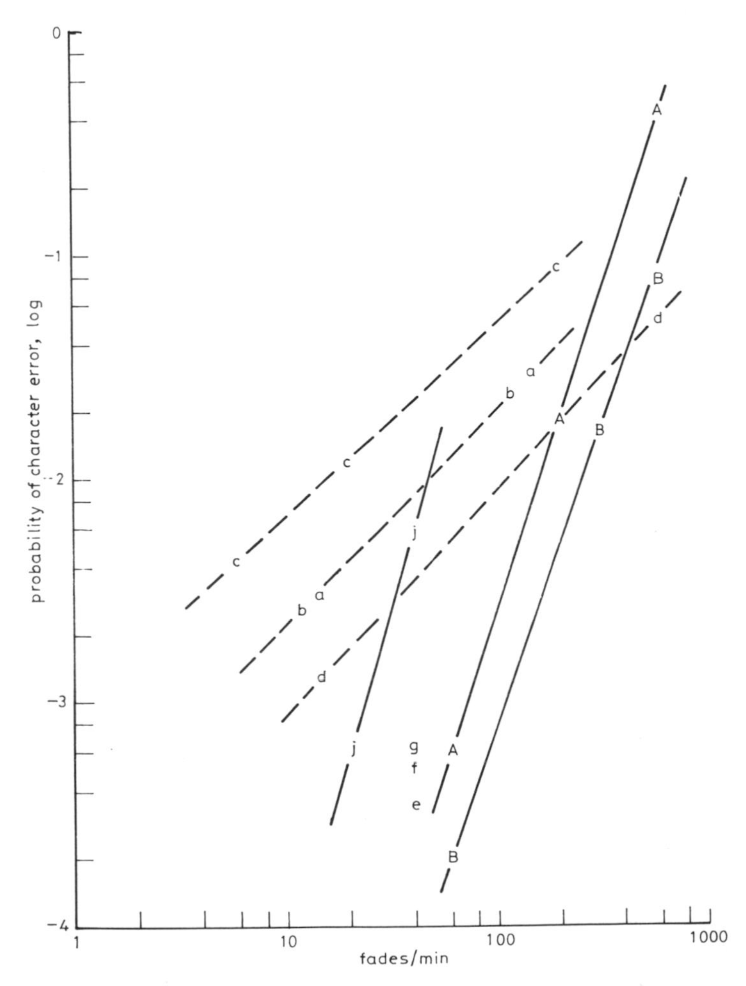

Fig. 7.4 *Character error rates (measured on channel simulators) for various binary and MFSK systems with fast fading signal*
(For system identification see Table 7.3)
Measurement conditions:
a, b, c, d, e, j : no noise
f, g : 40 dB s/n

shift is often associated with flutter fading and distorted conditions generally, there may be a limit on the useful information to be obtained from theoretical considerations and channel simulation.

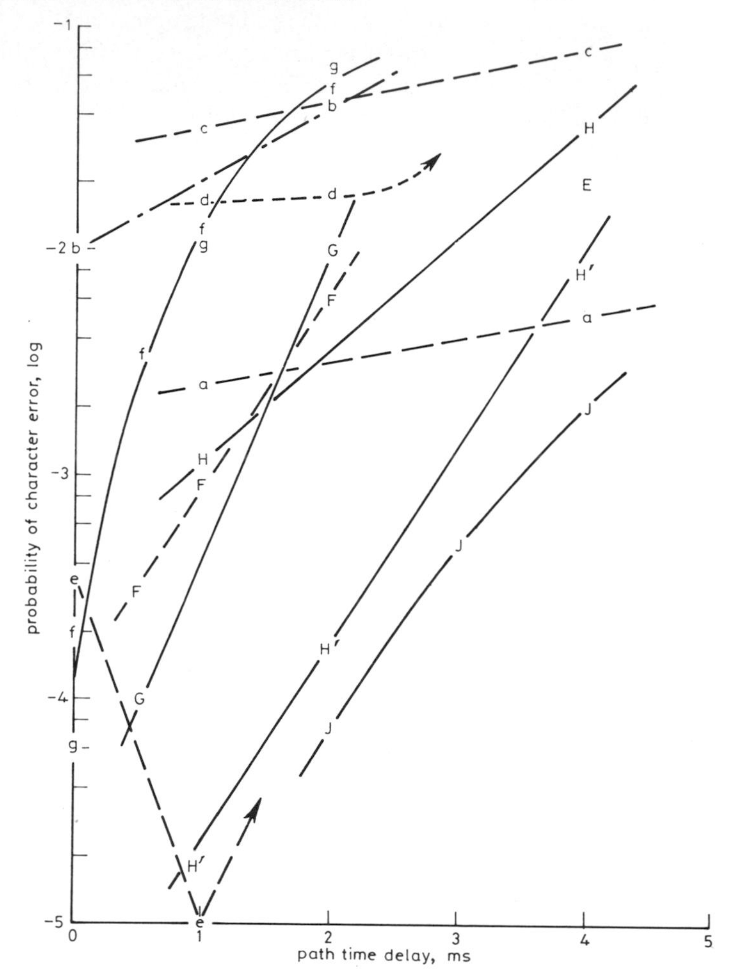

Fig. 7.5 *Character error rates (measured on channel simulators) for various binary and MFSK systems with multipath signals*

(For system identification see Table 7.3)

Measurement conditions:

a, c, d : 12 fades/min, no noise

j : 20 fades/min, 40dB s/n

e, f, g, h : 40 fades/min, 40dB s/n

b : 60 fades/min, no noise

When attempting to compare Kineplex, ANDEFT or any other multiplex system with MFSK, care should be taken to distinguish between characteristics which are fundamental to the method of modulation and those due to multiplexing. The measured performances given here are those of each single channel and so may be directly compared. However, a multiplexed system can not only

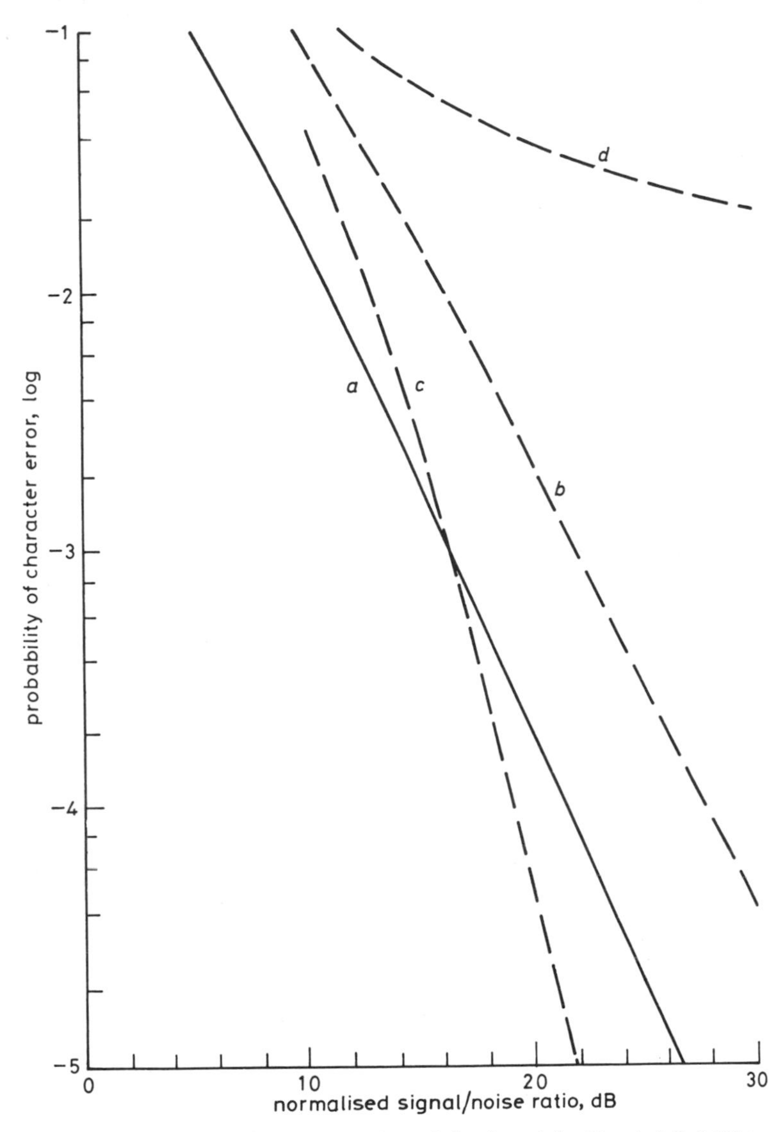

Fig. 7.6 *Character error rates (measured on channel simulators) for Piccolo Mk 1 (32 tones) and FSK with filter assessor detection (100 bauds, 510 Hz total shift, 5 × element errors)* [Reproduced by permission of British Telecom]

Fading 20 and 40 fades/min; dual diversity reception.

a Piccolo. Path time delays up to 4 ms

b FSK.; No delay and 2 ms PTD

c FSK.; 1 ms PTD (optimum for deviation)

d FSK.; 4 ms PTD

drastically reduce the normalised bandwidth, but the higher traffic capacity offers facilities for forward error correction and other techniques which must be evaluated separately. Some guidance on the comparisons of multiplexed systems is given in Section 7.7.

It may be appropriate to discuss at this point the possibility of a differentially-coherent MFSK system, i.e. one in which the matched filters are operated coherently and the audio phase of one element is extracted in the receiving system to provide a reference for the phases of the coherent matched filter detectors during the next element. Such a system is no doubt technically feasible, and, by analogy with PSK and DPSK, may give an improvement of up to 3 dB under slow fading conditions. However, fast-fading errors are a function of element length and are already a consideration on MFSK systems. To effectively double the time for which coherence is required will exaggerate this effect. There are also practical problems, such as the need to program into the system any frequency-dependent phase variations, which may be considerable under multipath conditions and vary between equipments. Altogether, the benefits of such a system seem dubious.

7.5 Non-coherent FSK detectors

The ideal non-coherent detection method for an FSK signal (system E) assumes orthogonal signals and detects each in a separate matched filter, comparing the detected outputs. As such, this is nominally identical to an MFSK system with $M=2$, and the curves C and E are coincident (with the reservation discussed in Section 7.3). Similar reasoning to that for PSK indicates that the performance on a fading signal is about 3 dB worse than coherent FSK and so 6 dB worse than coherent PSK.

There are two principal detection methods used in practice. In the first (referred to here as 'FSK limiter discriminator'), the signal is heavily amplitude-limited or zero-clipped (see Section 5.4) before being applied to a nominally linear frequency discriminator centred on the mean of the two frequencies. The polarity of the d.c. output is then assessed at the end of each element. This method, referred to variously as FSK or voice-frequency telegraphy, replaced on–off keying and was the major technique in machine telegraphy over h.f. for many years. As radio-frequency stabilities improved, the frequency shifts used were progressively reduced towards the optimum of $0.8B$ (where B is the modulation rate in bauds). A typical CCIR recommended format uses 80 or 85 Hz total shift, allowing 15 channels at 100 Bd to be frequency multiplexed in a nominal 3 kHz bandwidth [7] (see Table 7.2).

Channel simulator measurements [17, 18] with dual diversity reception show a DF of about 4 dB below a *coherent* FSK – about 1 dB below ideal non-coherent FSK. However, the system breaks down rapidly under multipath conditions, a PTD of 1 ms giving an element error rate greater than 0.2% (see f in Fig. 7.5). As would be expected from the wide bandwidth and the method of detection, the immunity to fast fading is quite good.

In 1956 an improved method of detection was developed [6], primarily to improve accuracy under multipath conditions. The system is often referred to as frequency-exchange keying or two-tone keying (from the methods of generation of the transmitted signal), but these titles are misleading, since, apart from the method of determining the optimum frequency shift, the major difference is in the method of detection of the signal, and it will be referred to here as 'FSK – filter assessor'. The two tone frequencies are separately filtered, the filter characteristics being designed to approximate to a matched-filter response, but without 'dumping'. The two detected outputs are subtracted and the polarity of the difference with reference to the long-term difference is assessed at the end of each element. Thus, in so far as the frequencies are orthogonal (or effectively so) and the filter characteristics effectively matched to the element, the performance should approach the ideal of system E, and measurements confirm this for flat fading.

However, the advantage of the system becomes apparent with multipath, if the delay is equal to half the inverse of the total deviation (e.g. 1 ms for a total shift of 500 Hz). Under these conditions, a selective fade which causes a null on one frequency will cause a maximum on the other, (see Section 4.6.2), and since the demodulation process can operate on either tone individually (effectively keyed on–off), there is very effective diversity operation, even with single-aerial reception. Measurements confirm that *single-aerial reception* with optimum PTD gives a DF of about 0.5 dB referred to system E *with dual diversity* – obviously a major improvement. As the multipath delay increases above the optimum, the performance should theoretically cycle between that for optimum delay and for flat fading at intervals of the optimum delay, but in fact for a 100 Bd signal, delays of about 4 ms or more produce the 'synchronising error' effect (see Section 4.6.1) and the system rapidly fails (see Figs. 7.5 and 7.6) [19].

In order to obtain the benefit of filter assessor operation, therefore, the deviation must be selected with reference to the current multipath delay. For a general-purpose system this means catering for PTD between 0.5 and 3 ms, ideally requiring shifts variable from 160 Hz to 1 kHz [19]. A compromise solution compatible with the multiplexed system referred to above is to use the centre frequencies (which are spaced at 170 Hz intervals), and select for each channel a pair of frequencies spaced approximately at the optimum [20] (the interleaving of frequencies for different channels being permissible). Although this system is no doubt effective for fixed point-to-point working of multiplexed links on large trunk routes, where the daily schedules are such that an optimum frequency pattern can be evolved over a period of time, or on which ionospheric sounding techniques are available, it is less practicable in a single-channel system, in emergency links, after changes in frequency allocation or when the characteristics of the ionosphere change (either as a result of short-term disturbances or in the annual or sun-spot cycle).

The measured data for multipath (Fig. 7.5) is confused and difficult to interpret for three reasons; much of the comparable data is lacking, the data available was measured under widely differing conditions, and some of the curves are periodic, as in Fig. 7.10 (note that the identifying letters on the curves indicate spot data and the interpolating lines are for guidance only).

For instance the FSK system with 85 Hz shift (curve c) would, from the basic theory of filter assessor operation, give a steady reduction in error rate as the multipath increases, to a minimum at about 6 ms. However, with a modulation rate of 75 Bd a synchronising error of this magnitude represents almost half an element duration, which would cause a major deterioration in performance. It is presumably the tendency towards mutual cancellation of these two effects which maintains a reasonable flat curve. The 850 Hz shift ('d') may be expected to give its optimum performance at delays of about 0.6 ms and 1.8 ms. The rapid breakdown for longer delays is typical of published measurements on FSK systems (Cf. Fig. 7.6) but since most of these are on 75 Bd or 100 Bd systems, this effect is probably due to timing errors rather than selective fading. Both of these curves are grossly pessimistic because of non-synchronous operation, and a better estimate of the absolute levels achievable is given by system 'e', noting particularly the improvement in performance with optimum delay (1 ms).

The normalised bandwidth of a single channel is dependent on the deviation, as indicated in Table 7.2, but in a multiplexed system is normally assumed to be twice that of a limiter discriminator system.

Although a number of comparison trials over radio circuits between MFSK and FSK techniques have been carried out by various authorities, of those available for publication only two were sufficiently rigorous to justify quoting quantitative results. Those carried out by the School of Signals as part of the tests mentioned above were particularly thorough, using standard military operational and engineering techniques over four different (short and medium) ranges, three different transmitter power levels and two different FSK equipments. The curves given in Fig. 7.7 are typical of the results which were consistently obtained on most tests.

Trials carried out by the author in co-operation with the Christien Michelsen Institute, of Bergen, Norway (as part of the work in support of COST 43, see Section 1.2) involved fixed frequency, round-the-clock transmissions every hour of a two-minute test message on the MFSK and FSK (filter assessor) systems in direct succession. Transmitter radiated power was very low and signals extremely weak, giving a good comparison of the two techniques under marginal conditions (see Table 7.4). Error coding was ignored as being equally applicable to either technique and both systems were accurately synchronised before each transmission.

Two trials of which only subjective or qualitative reports are available are none the less worth quoting. In 1972 SHAPE (Supreme Headquarters Allied Powers in Europe) Technical Centre at The Hague, Netherlands carried out an assessment of the Piccolo Mk 3 [21] (as code 1 of Table 6.1 and (a) of Table 7.3, with

Table 7.4 Summary of 'Bergen' radio trials

	FSK	MFSK
Number of Tones	2	10
Frequency interval	100	20 Hz
Element duration	16.67	50 ms
Modulation rate	60	20 elements/s
Data rate	60	66 bits/s
Data word	10	3 elements
Bandwidth	260	170 Hz

Synchronisation for each system adjusted accurately before each test.
Date of trials: 2nd–17th April, 1975
Radio link

Transmitter site:	Off-shore near Bergen, Norway
Aerial:	8 metre (?) whip
Power:	25 watts nominal
Frequency:	4.16 MHz
Receiving site:	Hanslope, Buckinghamshire, UK
Range:	1000 km approximately
Aerial 1:	4–8 MHz rhombic. Approximately 30° off bearing
Aerial 2:	Various (as available)

Synopsis of results

Total test runs measured	228
Number of runs with zero word errors in both systems	11
Number of runs with > 16% word errors in both systems	69
Number of runs analysed	148
Geometric mean (for all analysed runs) of:	
$\frac{\text{Word errors on binary system}}{\text{Word errors on MFSK system}}$	13.3

Availability (Defined as the percentage of *total* test runs giving word error rates less than acceptable: note no error coding):

Maximum acceptable word-error rate	1%	10%
Availability of binary system	9.7%	31%
Availability of MFSK system	32.3%	60%

continuous a.m. synchronising). This included observation of Piccolo reception over long-distance paths and direct comparison with a 75 Bd start–stop FSK filter assessor link over a shorter path. Comments include:

> . . . the individual Piccolo tones could barely be heard in a 600 Hz bandwidth (because of man-made background noise), yet throughout the demonstration consistently high-quality telegraphy copy was received . . .

> . . . the results indicated that the error rate of Piccolo was between one hundred and two hundred times lower than that of a start–stop system . . .

In 1977 the Swedish Material Administration carried out direct comparison trials between four different h.f. telegraphy modems:

32 tone Piccolo (LA 2028, see Section 1.2).

A synchronous 100 Bd FSK system with 170 Hz deviation and error indication system.

A double-frequency-shift system using two frequencies in succession for each state and a modulation rate of 150 Bd to give 75 bits/s.

A time/frequency-division multiplex DPSK equipment which sends data successively on each of 7 tone frequencies with a time interval between each.

The radio path was 250 km long in northern Sweden and suffered from heavy

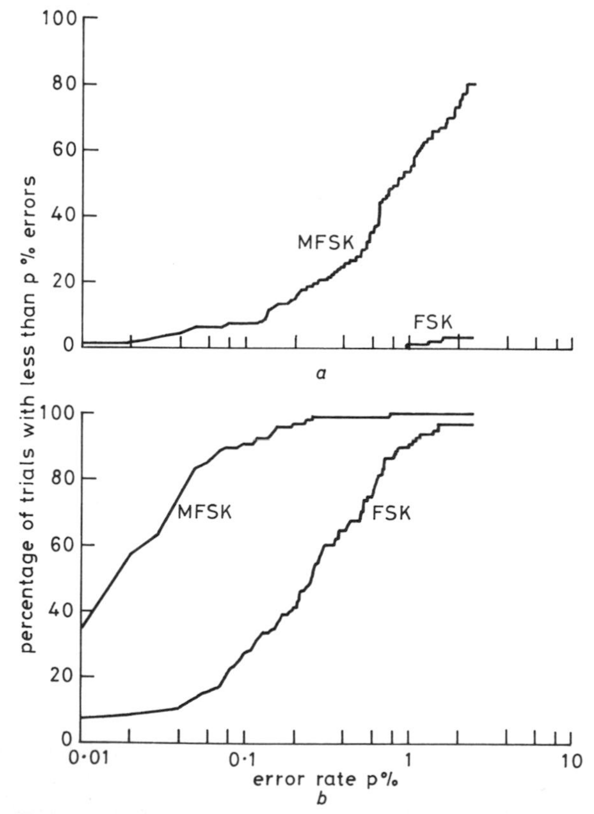

Fig. 7.7 *Character error rates (measured on comparison trials over 650 km radio path) for Piccolo LA 2028 (32 tones) and FSK with filter assessor detection.*
a Transmitter power 15 W; FSK shift 85 Hz total (c in Table 7.3)
b Transmitter power 125 W; FSK shift 850 Hz total (d in Table 7.3)

ionospheric disturbances from auroral discharges. The results (unpublished) indicate that Piccolo gave a higher availability than the other three systems, particularly when the propagation conditions were badly disturbed.

For a possible explanation as to why these and other undoubtedly successful trials produced little positive support for Piccolo, see Section 1.3.

Fig. 7.7 and the comments by SHAPE both suggest that in comparison with an equivalent FSK system, with start–stop (asynchronous) detection, an MFSK system should give an improvement of 100:1 or more in character error rate. The use of synchronous working for the binary system reduced this advantage to 10–15:1 (Table 7.4), but even this reduction still leaves a margin of two to three times the availability on weak signals. These results ignore error coding, and the use of such techniques enables raw data at quite a high error rate to be improved to a usable standard, although under these marginal conditions a relatively small deterioration in input errors can produce a pronounced deterioration in the output data (see Section 11.2).

The Swedish trials are of interest because the three non-MFSK systems represent in broad terms the techniques usually employed to try to improve marginal links, i.e. error coding, in-band frequency diversity, and combined time and frequency diversity. The last is a particularly powerful system often considered the ultimate in poor-signal working, although involving heavy penalties in normalised bandwidth, decoding delay etc. (see Sections 5.2.3 and 7.6.2). Even so, it is still weaker than a simple MFSK link (which, of course may be further improved by use of the auxiliary techniques).

7.6 Other Systems

As indicated in Section 7.1, consideration of practical systems is limited to those which may be identified with one of the theoretical principles analysed, and on which channel simulator measurements are available. This rules out a number of interesting experimental and specialised modem systems. However, in some cases limited conclusions may be drawn but it should be emphasised that these are based more on rule-of-thumb, experience and intuitive concepts than on detailed analysis or study of the systems concerned.

7.6.1 Chirp

'Chirp' binary telegraphy systems [22] use a 'gliding' tone (frequency modulation with a linear rate of change) with an increasing frequency symbolising 'mark' and decreasing 'space'. Detection is by a matched filter technique which produced a pulse output at the end of the sweep, of amplitude proportional to the correlation between the signal and the transmitted waveform. As a binary matched-filter technique using two filters and operating on a non-coherent signal, the signal-to-noise performance should be nominally the same as that of systems C and E in Table 7.1. Under multipath conditions a number of output pulses will be produced, separated in time by the multipath delay and therefore delayed signals may be discriminated against, or in principle integrated into a combined response without inter-element interference. This gives an improvement in performance against the time delay effects of multipath, but note that, since detection relies on the coherence of the signal over an element length, there is no corresponding protection against the phase perturbations due to fading nulls, and the element error rate under fast fading conditions (whether flat or selective) should be similar to that of an MFSK system with *the same element length*, and has the added disadvantages of the shorter element and wider bandwidth. Immunity to steady tone interference should be better than average but the wider bandwidth could give a weakness in the face of speech and similar 'random' spectra.

7.6.2 Adaptive Systems

The Kathryn system [23] is an extension of the ANDEFT system discussed above but with an adaptive facility which in theory allows the demodulation system to compensate for fluctuations in the communication path. Each tone carries not only a data modulation, but a pilot modulation known at the receiving terminal which may be separately extracted. Measurements carried out on this pilot modulation enable perturbations due to the ionosphere to be extracted and to be applied as a correction to the tones before demodulation of the data.

Two fundamental weaknesses in such adaptive systems carried out with analogue processing are, first, that (unless wide-band signal delay circuits are included) the adaptation must be delayed and averaged, and so will be unlikely to give an appreciable improvement in rapidly fluctuating conditions from any cause (flutter fading, severe multipath, impulse interference etc.) Secondly, the adaptive process is essentially multiplicative, so that if interference and noise added to the signal (largely rejected by narrowband linear processes) can contribute components to the adapting waveforms, these will modulate the signal itself and so deteriorate the demodulator performance.

Other adaptive techniques have been proposed, in which the (FSK or PSK) signal is analogue-to-digital converted and then processed by digital filtering and delay techniques programmed from information derived from a sounding pulse periodically interspersed in the data stream. Recursive feedback from after a threshold decision enables delayed signals to be cancelled [24]. While such techniques may be able to correct for specific and easily-definable effects, such as reasonably stable multipath conditions with low levels of noise and interference, even this requires quite complex processing. It would seem unlikely that a flexible general-purpose system able to deal with the wide variety of interference and ionospheric effects encountered in practice will be developed for some considerable time. It is difficult even to imagine an electronic circuit operating on a signal barely above atmospheric noise, and yet able to distinguish between, say, a signal fading at 10 fades/second and one which is suffering from interference from a carrier 10 Hz away.

Indeed, adaptive systems generally may well prove an illustration of the basic principle that a system developed to counter a specific effect may be found to give degraded performance against others.

7.6.3 Spread Spectrum

A technique known as spread spectrum modulation [25] spreads the telegraphy signal over a very wide band (possibly up to several MHz), either by multiplying the binary telegraph signal by a high-rate pseudo-random bit stream before r.f. modulation, or by 'hopping' the radio frequency at high speed over the complete band in a pseudo-random fashion. In either case the information is recovered in the receiving system by reversing the process, using an identical pseudo-random sequence which must be synchronised in time to the incoming transmission. Any non-synchronised signal, whether wide-band or narrowband is converted in this process to a noise-like interference, so that a number of transmissions of similar types or of totally different types may simultaneously occupy the same bands. Where similar systems are superimposed, the pseudo-random key streams of each system must be selected to prevent false synchronisation and mutual interference. These techniques were originally investigated to give security against jamming and eavesdropping to military tactical communications but have more recently been advocated for general use. Later suggestions involve some elements of time-discrimination, as in Chirp, and adaptive equalisation. The sophisticated wide-band transmitter and receiver techniques which are required and the necessity to synchronise a long, high-speed key stream with a time-varying and distorted signal suggests a level of electronic sophistication several orders higher than with other systems.

If the 'hopping' is at a higher rate that the element period, then the received signal is no longer coherent, even over an element, which must involve a serious loss of demodulator efficiency, and there are no positive indications of any improvement in performance against ionospheric effects (although, of course, an increase of immunity to narrowband interference). The arguments that the superposition of signals allows an acceptable bandwidth occupancy to be achieved is weak and based on unrealistic premises. Altogether, there must be very serious doubts as to the suitability of these techniques other than for very specialised applications, and even these must be assessed against their effect on other communicators, particularly in the heavily occupied h.f. bands.

7.7 Multiplexed Systems

7.7.1 Spectrum Economy

The earlier multiplexed FSK systems consisted merely of a number of single limiter-discriminator channels operated in parallel within the same receiver bandwidth. Since these systems were non-synchronous and non-orthogonal, the frequency separation between channels needed to be sufficient to ensure an acceptable level of inter-channel interference. To achieve high information density, therefore, required narrow deviations and highly selective filters. The minimum bandwidth of the filters and the deviation are both limited by the maximum modulation rate which may be transmitted and the frequency stability of the radio equipment.

The introduction of the filter assessor system (which allowed interleaving of frequencies for different channels) and synthesised radio equipment reduced some of these limitations but introduced others, as described above, so effectively doubling the normalised bandwidth.

The use of an orthogonal, synchronous system offers a major increase in bandwidth occupancy and Kineplex, ANDEFT and other DPSK modems are designed specifically for multiplex operation. If the modulation of the tones is synchronous on all channels (i.e. the times of transition of the frequency or phase changes coincide), and the frequency separation is correct then the signal in any one channel can be made orthogonal to the rest and in principle no filtering is required to separate the channels. Thus in the Kineplex and ANDEFT systems the tone separation is the inverse of the integration period and the composite signal is applied to all detector filters.

Although a true multiplexed Piccolo system has not yet been required (the Mark 6 Piccolo has some provisional multiplexing facility but is not an optimum design), there is no reason why such techniques should not been employed, and an MFSK system designed in which the separation between the extreme tones of two adjacent channels is made the same as the intertone separation within a channel. With such a system (discussed in Section 12.4) the normalised bandwidth is invariant and may be derived by a simple extension of the principles discussed in Section 3.4.4. The normalised bandwidths of MFSK and DPSK multiplex systems in Table 7.2 are calculated on this basis, but those for the FSK systems are derived as described above.

It is important to note that in many cases the design of a multiplex system is approached from a different standpoint from that of a single channel. The primary aim in designing a single channel system is usually to achieve the best possible performance (in terms of the lowest error rate under a given set of conditions) at a fixed data rate, and provided that the resulting bandwidth is acceptable to the available radio equipment, the bandwidth itself is only one of the factors which help to determine the performance (either directly in terms of the amount of noise reaching the detector system, or less predictably as governing the width of the minimum interference-free band in which communication can take place). To that extent the bandwidth is a secondary parameter resulting from the design. In the design of a multiplexed system the bandwidth is usually fixed (typically a 3 kHz allocated 'speech channel') and the aim is to get the maximum information density within this bandwidth while still achieving an acceptable error rate.

7.7.2 Comparison of Systems

This section will consider some of the factors which should be taken into account when comparing multiplexed systems, and show how they may affect the comparisons already made. It may be thought that consideration of demodulation factors and normalised bandwidths alone should provide sufficient data on which to compare the performance of two multiplexed systems but this approach can be misleading.

A point which may be overlooked is that the power rating of a telegraphy transmitter is based on the assumption that it is radiating a single unmodulated carrier, whereas an SSB transmitter is really 'voltage rated' in the sense that it requires a specific peak-to-peak modulating voltage input. Thus if two tones are simultaneously applied to the same transmitter, each must be half the specified voltage and therefore develops only one quarter of the rated output power. The total power output from the transmitter is therefore only half its rating. For simultaneous modulation by multiple tones, the power output per tone is reduced by the square of the number of tones.‡

As a practical example of comparison of modulation techniques in multiplex applications, the two compared here are:

(a) A 6 tone MFSK system as system B in Table 7.2, multiplexed to the required data rate by adding further channels, each with a frequency offset which is assumed to be orthogonal or sufficient to ensure no inter-channel interference.

(b) A two-phase DPSK system as system D in the table and with parameters complying with the specification MIL-STD-188C section 7.3 discussed in Section 7.4.1. This specifies 16 data carriers spaced at 110 Hz, each modulated at 75 Bd (13.3 ms). The quadrature modulation modes, and any additional pilot tones are ignored. For two-phase modulation the maximum total data rate is 1200 bits per second (although in the specification this speed is attained using quadrature modulation and in-band frequency diversity), and lower speeds are achieved by in-band frequency diversity.

It is assumed for analysis purposes that the modems for both systems are designed to accept two receiver signals in dual (space) diversity. The different systems which are compared are therefore as listed in Table 7.5 together with the theoretical performance figures when receiving a Rayleigh fading signal in Gaus-

---

‡ The sum of a number of sine waves of differing frequencies gives a waveform which is noise-like in having a high ratio of peak to RMS and some of the loss of efficiency may be regained by limiting the peaks. This causes waveform distortion, inter-modulation between various tones and other effects. The amount of improvement attainable for a given level of permissible performance deterioration is a complex subject which does not seem to have received the attention it would apparently warrant. Its application to MFSK multiplexing has not so far been considered and it is ignored in this analysis.

sian noise. The element error rates for the DPSK systems may be derived from the fundamental equation (Cf Table 7.1):

$$p = 0.5/(1+R)^D \simeq 0.5/R^D \qquad (7.1)$$

From this equation the signal-to-noise ratio R for different orders of diversity (maximal-ratio combining) is given by:

$D =$	1	2	4	8	16	32
R(dB) $=$	34.0	16.9	7.8	2.2	−2.0	−5.6

Table 7.5 *Theoretical transmitter power ratings for different multiplexed systems*

MFSK		DPSK (2-Phase)		Single aerial		Dual space div.	
Number of channels	Eff. data rate (bits/s)	Eff. data rate (bits/s)	In-band div. order	Total div. order	Transmitter power rating (dB)	Total div. order	Transmitter power rating (dB)
		1200*	1	1	29.1	2	29.1
		600	2	2	12.0	4	20.0
		300	4	4	2.9	8	14.4
6	300			1	15.5	2	15.5
5	250			1	13.9	2	13.9
4	200			1	12.0	2	12.0
		150	8	8	−2.7	16	10.2
3	150			1	9.5	2	9.5
2	100			1	6.0	2	6.0
		75	16	16	−6.9	32	6.6
1†	50			1	0†	2	0†

*Not available in MIL Spec. †Reference system

The performance of the MFSK systems can be derived from the equations given in Chapter 3. However, the comparative performance of any two of the systems may be obtained more simply by starting from the demodulation factors listed in Table 7.2 (which compares equivalent basic systems) and then applying 'correction factors' to arrive at the relative performance, expressed in terms of the *rated* transmitter power required to maintain the same mean error rate. The various factors are derived as follows:

(a) A single 6-tone MFSK channel sending 50/bits/s and radiating the total rated transmitter power is taken as the reference system (0 dB), whether received single-aerial or dual space diversity.

(b) For an 'n' channel frequency multiplexed signal without clipping, each data tone must be $(1/n)$ of the required total amplitude, e.g. for $n=6$ the rated transmitter power must be +15.5 dB and for the 16 data carriers of the DPSK system the correction factor is +24 dB.

(c) the demodulation factor of DPSK is 0.5 dB with reference to the 6-tone MFSK system so that for an ideal DPSK system at 50 bits/s the transmitter power must be +0.5 dB.

(d) A single DPSK channel as specified operates at 75 bits/s so to maintain the same error rate as a 50 bits/s system would require an increase in transmitter power of 1.5:1 = +1.76 dB.

(e) As previously discussed, the DPSK system has a guard band of 4 ms, so that approximately 4/13 of the transmitter power is wasted, requiring an increase of +1.75 dB.

(f) The order of in-band diversity used in the DPSK systems requires a correction factor derived from the table above. When considering single-aerial reception the factor is the difference in dB between the in-band diversity mode and non-diversity, while for dual space diversity reception the correction factor is the difference between twice the in-band order and dual diversity.

Thus for dual space diversity reception, a 6-channel 6-tone MFSK system requires a rated transmitter output of +15.5 dB while the equivalent 300 bits/s DPSK system (which uses 4-fold in-band diversity), requires (+.5+1.76+1.75+24 +2.2−16.9) = +13.3 dB.

From the table the basic characteristics can be seen. Starting from the single channel MFSK reference, as the number of MFSK channels is increased the transmitter power must be increased proportionally to the square of the number of channels. The basic 1200 bit/s 2-phase DPSK system requires about 30 dB more power than the reference but as the data rate is reduced by diversity combination the transmitter power required is reduced. When operating with a single aerial the MFSK is obviously at a disadvantage in having no diversity at all and so the DPSK system shows an advantage of 6 to 12 dB. However, when both systems are operating in dual space diversity the advantage of the higher orders of diversity in DPSK is considerably less, so that the systems are roughly equivalent at 300 bits/s and the MFSK some 4 dB better at the lowest rates.

Unfortunately there is a serious weakness in this theory as applied to the DPSK modem. The calculations on diversity performance assume that all the signals being combined are uncorrelated in fading pattern. While this should be true for space diversity, it is not true for in-band frequency diversity reception of flat fading, and since even under multipath conditions there will be a measure of flat fading on each of the multipath components separately, the figures for the multiple in-band diversity cases must be considered to be very optimistic and the calculated advantage of DPSK under these conditions is not realised in practice. Measurements made in the author's organisation on a commercial DPSK system complying with the MIL specification in the 300 bits/s and 600 bits/s modes, are shown in Figs. 7.8, 7.9 and 7.10. The curve for the MFSK system in Fig. 7.8 was derived from single channel measurements, corrected as in (b) above, and shows an improvement of 6 dB over the directly comparable DPSK system.

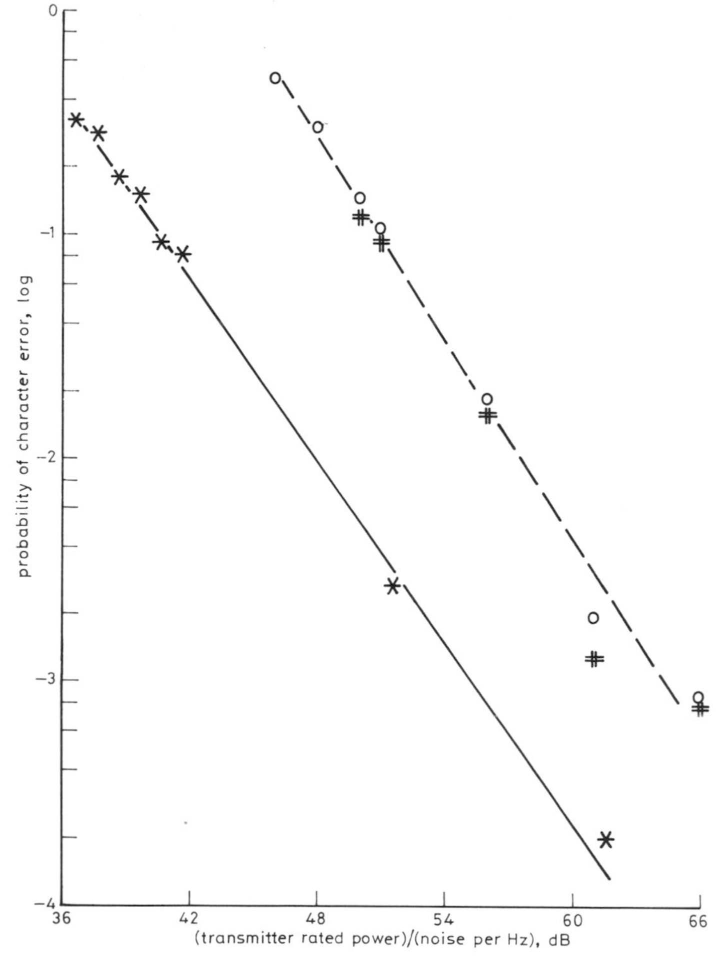

Fig. 7.8 *Character error rates (measured on channel simulator) for multiplexed systems*
Fading 12 fades/min; dual ('space') diversity reception
* Piccolo Mk 6 (6 tones). 6 channels multiplexed; 300 bits/s total
○ DPSK 300 bits/s (four-fold in-band time-division multiplexed)
⧺ DPSK 600 bits/s (dual in-band time-division multiplexed)
DPSK character errors = 5 × element errors

In the fading-in-noise and fast-fading situations there was no significant difference in measured performance between the two DPSK systems. In the multipath situation the expected results are obtained, the minima and maxima of the error rate corresponding roughly to the time delays predicted from the basic theory of Section 4.7.2 as indicated on the diagram. These results strongly support the conclusions drawn from theory, that in-band frequency diversity can only improve the error rate to the extent that errors are due to multipath reception with specific time delays and that the predicted advantage is not realised in the general case.

It is evident that in-band frequency diversity is often an inefficient method of improving data quality by sacrificing normalised bandwidth. Other and preferable methods are discussed in the Appendix.

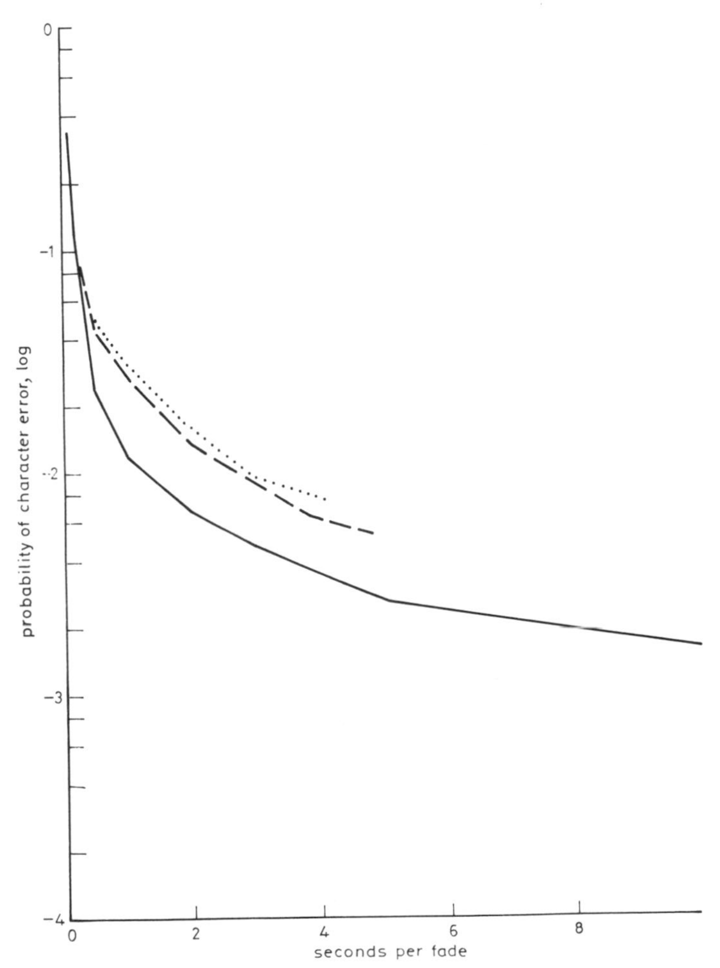

Fig. 7.9 *Character error rates for multiplexed systems with fast fading signals*
——— Piccolo Mk 6
- - - - - - DPSK 300 bits/s see Fig. 7.8
. . . . . . DPSK 600 bits/s

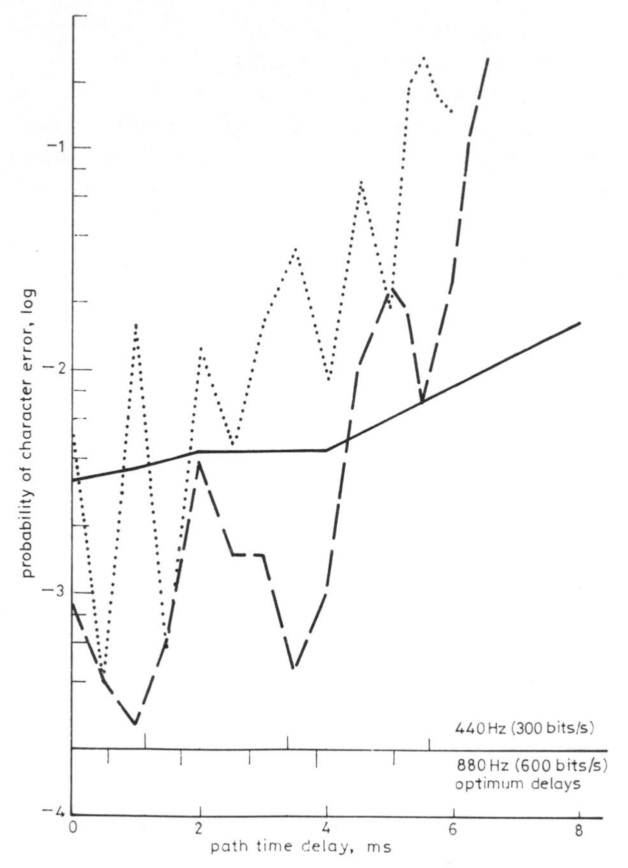

Fig. 7.10 *Character error rates for multiplexed systems with multipath signals*
See Figs. 7.8 and 7.9

7.8 Discussion and Summary

Since the measured DFs of each system under slow flat-fading conditions are within about 1 dB of the calculated performance of that system, the relative performance of practical equipments may be predicted from Figs. 7.2 and 7.3 (with appropriate corrections for data rates, synchronising etc.). It can be accepted that a 6-tone MFSK system will perform better (by about 4 dB) than an equivalent DPSK system. Where a binary system uses an inefficient code format or non-synchronous detection the margin will be correspondingly greater.

The data on fast fading for binary systems is limited (Fig. 7.4), but suggests that the theoretical weakness of MFSK is not borne out in practice. For fading rates higher than about 30 fades/min (2 secs/fade), MFSK is about equivalent to or better than DPSK and narrow-deviation filter assessor, but worse than limiter-discriminator and wide-deviation filter assessor. Selection diversity will allow operation of MFSK with fade rates up to about 20% of the element rate (with error control if necessary), but for operation under limiting conditions such as this multiple frequency spacing or interleaving should be considered (see Section 12.1.1 and note that this technique would be ineffective with DPSK).

The filter-assessor system has a considerable immunity to the selective-fading effect of multipath, provided that the deviation may be selected to be optimum. Where this is not practicable then (as with DPSK and MFSK), the selective fading pattern is merely superimposed on the flat fading and the relative performances will be roughly according to Figs. 7.2 and 7.3. It is against the timing-error effect of multipath that the superiority of MFSK (arising from its longer element length) is most evident. While for MFSK the error rates rise slowly as the delay increases (even at very long delays), those for all binary systems rise rapidly once the delay (less any guard time) exceeds a few per cent of the element length. A detailed discussion of the MFSK measurements is given in Section 6.5.

The above discussion may be summarised in the following general conclusions:

(a) Where the highest possible information density is required in a wide-band multiplexed system, DPSK systems have a definite advantage, provided that use is limited to links engineered for high signal-to-noise ratios, and extreme ionospheric conditions are not encountered. Where these conditions cannot be satisfied, the higher error rate means that a lower data rate must be accepted, and care must then be taken to use effective methods of channel combining and data rate reduction. There is some justification for suspecting that in many practical links the higher traffic capacity offered by DPSK techniques is often squandered on unnecessary redundancy and inefficient combining methods. The assumption inherent in some modern equipment designs that it is advantageous to use the full allocated bandwidth even for lower data rates, is not only unjustified in theory and in practice but is another factor contributing to the overcrowding of the h.f. bands.

(b) The measured performance of DPSK systems on fast fading is disappointing (having regard to the relatively wide bandwidth of the matched filters), and the use of a guard band against multipath would seem to be essential, even though this sacrifices some of the performance in noise.

(c) The limiter discriminator method of detecting FSK can achieve a narrow normalised bandwidth, particularly in a multiplexed system but otherwise has little to recommend it, being particularly prone to multipath errors.

(d) The filter assessor method of FSK detection is a decided improvement, but in order to achieve its theoretical advantage against multipath, not only must the normalised bandwidth be considerably increased but some engineering, operational and system flexibility must be available to change the deviation according to the current multipath delay, which must be assessed by some means. Multipath delays which are long enough to enable the bandwidth to be reduced cause deterioration due to the variable timing errors they produce.

(e) An initial advantage of the MFSK principle is that for many applications it offers a choice of parameters to achieve a given data rate, and therefore may allow a better compromise with the effects of the communication channel. The theory would suggest a weakness under fast fading conditions but in practice the performance in this respect is comparable to the binary systems and very much reduced by diversity operation.

(f) The predominant advantages of MFSK for h.f. communication lie in the excellent performance in the face of noise (Gaussian and impulsive), speech and other wide-band interference, and multipath. It is probably true to say that when the error rate of an MFSK link rises to an unacceptable level because of ionospheric conditions or wide-band interference, no other modulation system with comparable overall data rate and bandwidth would be able to improve on the performance.

(g) The disadvantages of MFSK lie in the higher demands made on the stability of the radio system, the increase in electronic complexity and the loss of flexibility inherent in a synchronous system signalling in predetermined bytes, but all these limitations may be reduced to acceptable levels by the use of modern technology.

(h) Techniques such as diversity reception and error coding which are applicable to binary systems are also applicable to MFSK systems and improvements of the same order may be obtained (see Chapter 11).

(i) As a matter of general philosophy, progress in the improvement of communication over such a complex and variable medium as the ionosphere is unlikely to be achieved by the development of a number of new systems, each of which is intended to counter a specific problem. Any improvement against a single error-causing mechanism must be attained *without any loss of performance in other respects.* To win a single battle on a narrow front is a pointless victory if it involves retreat elsewhere.

As a general comment, it should be emphasised that if a signalling channel is unsatisfactory because of multipath, Doppler, fast fading and other effects which are not noise related, an increase in transmitter power will not in itself produce an improvement. The widely held belief that the performance of a weak radio link will usually be improved by the use of higher transmitter powers is not only often incorrect but is one of the factors contributing to overcrowding and high interference levels on the h.f. bands.

7.9 References

1 CCIR: 'Ionospheric channel simulators', Kyoto, 1978, Rep. 549-1, **III**, p.47

2 CCIR: 'Use of high frequency ionospheric channel simulators', Kyoto, 1978, Rec. 520, **III**, p.46

3 SCHWARTZ, M., BENNETT, W. R. and STEIN, S.: *Communication Systems and Techniques* (McGraw-Hill, 1966), Chapter 8

4 CCIR: 'Prediction of the performance of telegraph systems in terms of bandwidth and signal-to-noise ratio in complete systems', Kyoto, 1978, Rep. 195, **III**, p. 120

5 LAW, H. B.: 'The detectability of fading radiotelegraphy signals in noise', *Proc. IEE*, 1957, **104B**
6 ALLNUTT, J. W., JONES, E. D. and LAW, H. B.: 'frequency diversity in the reception of selectively fading binary frequency modulated signals', *Proc. IEE,* 1957, **104 B**
7 CCIR: 'Arrangement of voice frequency telegraph channels working at a modulation rate of 100 bauds over H.F. radio circuits', Kyoto, 1978, Recommendation 436–2, 3, p. 159
8 CCIR: 'Bandwidth of radiotelegraph emissions types A1 and F1', Kyoto, 1978, Report 179-1, **I**, p. 263ff
9 'Assessment trial of Piccolo telegraph modem', Army School of Signals Assessment Trial Report 248/78, Feb. 1978
10 Schwartz *et al.* As Reference 3, p. 445
11 de BUDA, R.: 'Coherent demodulation of FSK', *IEEE Trans. Comm.,* June 1972, p. 429
12 DOELZ, M. L., HEALD, E. T. and MARTIN, D. L.: 'Binary data transmission techniques for linear systems', *Proc. IRE,* May 1967, **45**, pp. 656–661
13 BRAYER, K. (ed): *Data Communication via Fading Channels* (IEEE Press 1975) (Includes papers and bibliographies on ANDEFT, Kathryn, Kineplex etc).
14 Military Communication Technical Standards. MIL-STD-188C Section 7.3.5 USA Dept of Defence
15 PORTER, G. C., GRAY, M. B., and PERKETT, C. E.: 'A frequency-differential phase shift keyed digital data modem', Report ESD-TE-66–639, Director of Aerospace Instrumentation, USAF 1966
16 Brayer. As Ref 13 Chap 7. (pp. 318–326)
17 RIDOUT, P. N. and WHEELER, L. K.: 'Choice of multi-channel telegraphy systems for use on h.f. radio links', *Proc. IEE,* 1963, **110**, (8), pp. 1402–1410
18 CCIR: 'Performance of telegraph systems on h.f. radio circuits', Report 345–2, **III**, p. 162
19 CCIR: 'Measurements of path-time delay differences and their incidence on radio circuits in case of multipath propagation', Kyoto, 1978, Report 203, 3, pp. 13–15
20 WHEELER, L. K., and LAW, H. B.: 'The performance of a 32-tone telegraph system subject to noise and fading', GPO Research Report No. 20879 March 1964 (now British Telecom)
21 SCHEMEL, R. E.: 'An assessment of Piccolo, a 32-tone telegraph system', SHAPE Technical Centre Memorandum STC TM-337, 1972, File Ref. 9980, DRIC No. P186242
22 GOFF, G. F. and NEWSOME, J. P.: 'H.F. data transmission using chirp signals', *Proc. IEE,* 1971, **118**, (9), pp. 1162–1166
23 Brayer. As Ref 13 Chap 7. (pp. 327–340 and 341–351)
24 HODGKISS, W. and TURNER, L. F.: 'Practical equalization and synchronisation strategies for use in serial data transmission over h.f. channels', *The Radio and Electronic Engineer*, April 1983, **53**, No. 4, pp. 141–146
25 CCIR: 'Spread spectrum techniques', Kyoto, 1978, Rep. 651, **1**, p. 4ff

8

Synchronisation

8.1 Introduction

In Section 2.6 synchronisation is discussed in terms of the process in the receiving equipment whereby a locally generated clock waveform at the element rate is adjusted in phase so that a specific point on the waveform (usually a transition) coincides as accurately as possible with the effective instant at which the frequency of the received signal changes between one element and the next. MFSK systems are essentially synchronous, and ideally the stability of the element repetition frequency in the transmitted signal and the intrinsic stability of the receiving system should be such that neither of these factors contributes to the synchronising error. The frequency tolerances required in order to achieve this are discussed in Section 10.4.2, and if these are met, we may assume that synchronisation errors are due to:

(a) variations in the time of arrival of the signal because of variations of signal path transit time;
(b) variations of delay in the sending and receiving equipments;
(c) inaccuracies in the synchronising system itself.

A particular case of a combination of (a) and (c) is when the synchronising system is too slow (or makes no attempt) to follow time-varying path delays and therefore leaves a residual time-varying synchronising error.

Delay variations in the equipment are usually negligible, with one exception. The communication path within the equipment (and particularly in the radio receiver) invariably includes bandpass filters with steep cut-off to exclude as much unwanted noise and interference as possible. Since these filters must also reject to some extent some of the sidebands created by the keying of the signal, it follows that the waveform after each frequency transition is correspondingly distorted, the effective instant of transition being delayed as illustrated in Fig. 8.1 (referred to as the 'group delay'). The amount of delay will vary according to the frequencies before and after the transition, and the design of the filter. A transition between frequencies near to the middle of the signalling band will usually be subject to slightly less delay than one between two frequencies near to the edges of the band. If the system has been accurately synchronised to the former signal, there will be a small synchronising error in the latter case. The spread of group delay over the signalling band should be checked for any radio receiver intended for use over MFSK circuits, and must be taken into account in the design of any audio band-limiting filters.

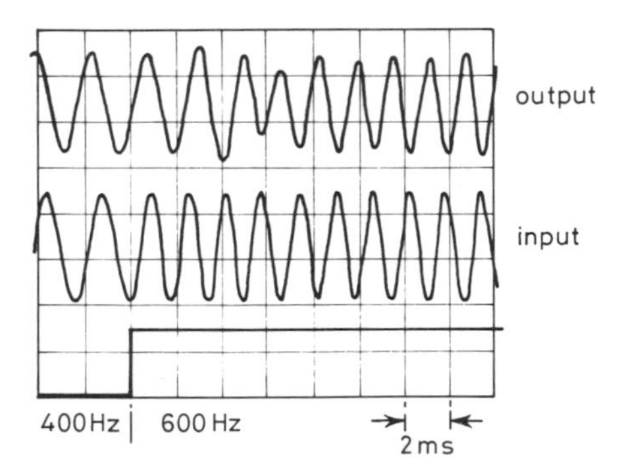

Fig. 8.1 *Oscillograms of frequency transition before and after bandpass filter, showing 'group delay' distortion*

In the design of a synchronising circuit, the objective of the modem designer should be to develop a system in which the synchronising error will never be such as to produce an appreciable increase in the intrinsic error rate due to the demodulation process itself. To do this requires an overall accuracy of a few per cent of the element length to be maintained on signals which may be several decibels below the level of noise in the signalling bandwidth, subject to high levels of interference of short duration, and fading at a variable rate. It is also appropriate to emphasise yet again that a signal which is transmitted as a continuous sine wave will, by the fading process in the ionosphere, be subject to phase variations as discussed in Section 4.4. In the author's opinion, the design of an adequate and satisfactory synchronising system is the most difficult single technical problem in the implementation of an MFSK system (and indeed many other systems), and he believes that the failure of many communication systems to achieve in practice the performance predicted by theory is all too often due to design weaknesses in this area. Many otherwise excellent books on communication theory tend to ignore such problems and (possibly for that reason) they are often ignored or underestimated by equipment designers, and in laboratory measurements of equipment performance.

There are obviously a number of ways in which synchronising can be achieved but the first consideration is the method by which the information on which the adjustment depends is conveyed between the transmitting and receiving systems. In general there are four possibilities:

Intrinsic synchronisation: the synchronisation process operates by examining the data signal itself.
Continuous synchronisation: specific synchronising information is conveyed simultaneously with the data signal (for instance as an additional sub-modulation).
Periodic synchronisation: synchronisation information is inserted into gaps in the data signal at regular intervals, synchronism being maintained during data by the stability of the send and receive rates.
One-shot synchronisation: synchronisation information is sent at the beginning of a transmission before the data, and is maintained by clock stabilities throughout the duration of the data stream.

8.2. Intrinsic Synchronisation

It is tempting to suggest that element synchronising information for MFSK signalling could be derived from the data signal itself (as is often done in binary signalling), and the author has made several attempts to develop a system using the frequency changes between elements to define the time of such change and has so far been unsuccessful. One of the fundamental problems is that the information available is relatively ill defined. The frequency change may be anything between 1 and (M-1) increments of frequency, positive or negative, or may be zero, and there is no step change of phase. An ideal frequency discriminator would, in theory, generate a stepped waveform with transitions at the element rate, and a step height randomly selected between $\pm(M-1)$. In fact consideration of practical parameters shows that only a very poor approximation to this waveform can be obtained (see Section 3.7), and a circuit to derive the transition times from this waveform with any degree of accuracy under noise conditions would seem to be difficult if not impracticable. Following an alternative line of thought, that the matched filters themselves constitute the best possible identification of the frequency of the signal, the waveforms of the responses of a series of matched filters under conditions of synchronising error have been studied, to see if there was some factor which could be extracted to give a sense, and preferably a proportional error signal, to enable correction to be made. This also has proved fruitless and the idea of intrinsic synchronisation has been abandoned.

8.3 Continuous Synchronisation

The data are conveyed by frequency modulation of tone of constant amplitude, therefore it is practicable to apply a small amplitude modulation simultaneously to the signal to convey synchronising information. This is the method used in the Piccolos Marks 1, 2 and 3. The element length in these is 100 ms, thus the signal is amplitude modulated 10% with a 10 Hz square wave. In the receiving system this modulation is extracted by a peak detector and the local clock pulled into phase with the extracted signal by a phase-locked loop. Over the years in which the equipment has been in use this system has been proved quite effective but it suffers from a number of disadvantages:

(a) The effect of the amplitude modulation on the signal is to distort the matched filter response waveforms so that the system is not perfectly orthogonal. For about 10% square wave modulation the deterioration can be calculated as being roughly equivalent to a worsening of the signal-to-noise ratio by about 1 dB.
(b) Because of the poor signal-to-noise ratio available from the amplitude detector it is necessary to restrict the bandwidth of the control loop, causing operation to be sluggish and initial pull-in from a large error to take up to 15–25 seconds.
(c) The amplitude modulation at 10 Hz is awkward to apply to class C telegraphy transmitters.
(d) The AGC time constant in the radio receiver is usually shorter than a half-period of the waveform and therefore causes envelope distortion. It is necessary to modify most radio receivers to give time constants considerably longer than the modulation period.
(e) Under some conditions there is a systematic effect which generates an appreciable timing error in one direction .

The last point is an extremely subtle one which was only discovered after detailed investigation. Severe multipath conditions (or poor frequency response in audio circuits) can cause large differential attenuation between different tone frequencies, which in effect amplitude modulates the received signal with a quasi-random step waveform, changing the level every 100 ms. Such a waveform has a considerable 5 Hz content which is fed back on the AGC control, amplitude modulating the signal. The effect of modulating a signal with 5 Hz, when the envelope of the signal already has a strong 5 Hz component, is in this case to produce a spurious amplitude modulation at 10 Hz, which happens to be almost 90° out of phase with the correct synchronising signal. Once identified, this effect can be minimised by better circuit design but may be taken as an example of the difficulties referred to in Section 8.1.

The apparent advantage of a continuous synchronising system is that it will follow drifts or changes in the sending data rate, but in fact rapid or erratic drifts cannot be followed because of the sluggish response of the control loop, so that control may be lost during a period of bad signal and a synchronising error created, which can take an appreciable time to correct when the signal improves. Since modern technology has simplified the provision of highly stable data clocks, there seems little point in accepting such disadvantages, and while a continuous synchronising system may be viable in specialised cases, the methods discussed below are preferred for h.f. telegraphy.

8.4 Periodic Synchronisation

It is possible to interpolate a synchronising signal at intervals in the data stream, in which case the signal may be of the same type of modulation as the data itself and many of the problems of sub-modulation are eliminated. For instance, if an additional frequency is allocated for a synchronising tone, an element at this frequency may be inserted in the data stream at regular intervals of, say, 10 or 20 data elements. The timing is adjusted until the output of a matched filter at this frequency is optimum.

The disadvantages of such a system are that, since the control information is discontinuous, the response must be relatively slow (requiring several repetitions of the synchronising element to pull-in accurately), the insertion of the synchronising element(s) must be taken into account in determining the data rate, as in Section 2.5, and in effect once the system has synchronised this percentage of the signal power is wasted.

8.5 Single-Shot Synchronisation

Consideration of the disadvantages of the above systems suggests that it may be advantageous not to attempt to control synchronism continuously during a data stream but to synchronise completely and accurately before a message begins and to rely on the stability of the sending and receiving clocks to maintain synchronism with sufficient accuracy for the full length of the longest continuous message to be sent. This method has considerable advantages, notably:

1 The synchronising signal may be similar in form and modulation to the data signal, requiring no additional sub-modulation.
2 The full signal power is available for synchronisation purposes, enabling a control signal to be derived with the maximum possible signal-to-noise ratio.
3 There are no constraints on the response time of the circuit. In principle any required degree of accuracy can be obtained by using a slower loop and the only penalty is the length of time for which a synchronising signal must be transmitted before a message begins.
4 The synchronising signal does not affect the data rate or data signalling process in any way.

The main disadvantages of the method are:

1 Variations of the absolute transit time delay during a message (Section 8.1, (a)) must be small and acceptable.
2 Additional tone frequencies may be required, thus increasing the system bandwidth.
3 The stability of the data rate clocks at the sending and receiving equipment must be sufficient to hold synchronisation to the required degree of accuracy for the full duration of the longest expected message (see Section 10.4).

Since there is a requirement for high frequency stability of the audio tones, the provision of stable data rate clocks does not constitute an additional problem because audio tones, matched filter reference frequencies, and the data rate clocks can all be derived by synthesis from a single reference crystal oscillator (see Chapter 9). The problem of re-timing a random or low-stability data input is discussed in Chapter 10.

Since the synchronising signal is transmitted before data is sent, there is no point in having a distinct 'idle' or 'standby' signal. In other words, the requirement is fulfilled completely if the standby signal also carries the synchronising information. The receiving system will then resynchronise if necessary on any break in the input data stream when the sending end returns to the standby condition.

8.6 The Synchronising Signal

If the system is to be 'code transparent' (in the sense that all symbols of the input data are treated identically within the modem system) the standby signal constitutes an increase in the input alphabet, as defined in Section 2.5. If the system uses a single element per character, it would be possible to add a single tone and to phase or frequency modulate it at element rate, but the system is slightly simplified if two additional frequencies are used, sending alternate elements of each, the modulation then being at half element rate. Where the system sends two or more elements per character, the synchronising has a subsidiary role in identifying which is the first element in a character (character synchronisation) and therefore it is sufficient to increase the alphabet by one character, ensuring that the coding is such that standby is not conveyed by the same tone sent twice in succession. With the further restriction mentioned in Section 6.6 (i.e. the three sequences closely resembling the standby signal are not used), for a single-element system an increase in alphabet of two characters is required, and for a two-element system an increase of four characters. Where the basic system already includes sufficient redundant characters (see Tables 2.1 and 5.1), the additional synchronising information does not require any increase in tones or bandwidth.

The position of the synchronising tone pair within the series of tones used for signalling is not particularly critical but the arrangement of placing them in the middle of the tone series is not only convenient, but it also means that the differential group delay between the synchronising tones and the data tones is a minimum for most of the tones, rising to a maximum only at the extreme frequencies.

The following paragraphs discuss such a synchronising system developed for the Piccolo Mark 6, in which the synchronising signal consists of the frequencies 500 and 520 Hz repeated alternately in elements of 50 ms duration. This is treated as a 'carrier' frequency of 510 Hz, phase modulated $\pm 90°$ with a 10 Hz triangular waveform as shown in Fig. 8.2. This is a neat, convenient and efficient arrangement in that the use of adjacent tones according to the orthogonal relationship described in Chapter 2 automatically gives a synchronising signal with the narrowest practical bandwidth compatible with the rest of the system, and a phase deviation which is the largest that can easily be extracted with a simple phase-locked loop and which will give the best signal-to-noise ratio.

8.7 First Phase-Locked Loop

The phase modulation envelope is extracted in a second-order phase-locked loop (PLL) with a lag filter, as indicated in Fig. 8.3.

Assume a sinusoidal phase modulation:

$$E(t) = \sin(\omega_c t + \theta_i \sin \omega_m t) \tag{8.1}$$

The reference frequency is varied by the voltage $V_1(t)$ so that:

$$E_r(t) = \sin(\omega_c + K_2 V_1(t))\, t \tag{8.2}$$

where K_2 is the sensitivity of the frequency control (rad/s/V).

The 'error' voltage $V_o(t)$ is assumed proportional to the phase difference between the detector inputs:

$$V_o(t) = K_1 \left[\theta_i \sin \omega_m t - \int K_2 V_1(t)\, dt \right] \tag{8.3}$$

where K_1 is the sensitivity of the phase detector (V/rad).

The transfer function of the filter is:

$$\frac{V_1(t)}{V_o(t)} = \frac{1 + j\omega_m \tau_1}{1 + j\omega_m \tau_2} \tag{8.4}$$

The analysis of such a loop is given by Moschytz [1] and others who derive the relationship:

$$\frac{\theta_r(t)}{\theta_i} = G_r \sin(\omega_m t + \phi_r) \tag{8.5}$$

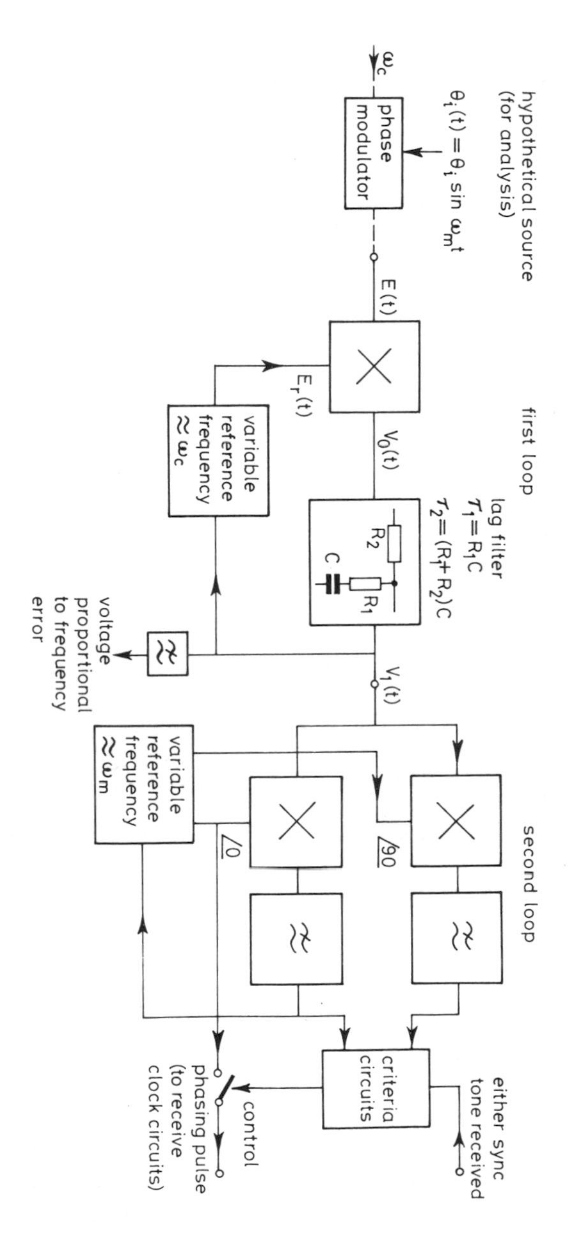

Fig. 8.3 *Block schematic of synchronising system*

However, the output for present purposes is the voltage after the filter, and:

$$\frac{V_1(t)}{\theta_i} = \frac{\omega_m}{K_2} G_r \cos(\omega_m t + \phi_r) \tag{8.6}$$

Note that while θ_i and θ_r are at the carrier frequency ω_c, ϕ_r is at the modulation frequency ω_m. It can be shown that the response is that of a single-pole bandpass filter and so may be conveniently expresses as:

$$\frac{V_1}{\theta_i} = \frac{\omega_o/K_2}{\sqrt{\{4D^2+(1/X-X)^2\}}} \tag{8.7}$$

$$\tan(\phi_r + \pi/2) = (1/X-X)/2D \tag{8.8}$$

where

K is the loop gain (in rad/s/rad)	$= K_1K_2$
ω_o is the 'resonant frequency' of the loop	$= \sqrt{(K/\tau_2)}$
D is the 'damping ratio' of the loop	$= (1+K\tau_1)/2\omega_o\tau_2$
X is the normalised frequency	$= \omega_m/\omega_o$

Curves for eqns, 8.7 and 8.8 are given in Fig. 8.4.

It is obviously advantageous to make $\omega_o = \omega_m$ for the synchronising signal (i.e. to 'tune' the loop response to half the element frequency) in which case the phase shift through the loop is zero and the only design parameter to be decided is D. An important consideration in the choice of D is that it is a function of the loop gain K which varies with the amplitude of the signal fed to the phase detector and therefore from eqn. 8.8 the phase shift through the loop will vary with signal amplitude, producing a variable synchronising error on a fading signal. A low value of D would give a narrow noise bandwidth and other advantages, but a lower limit is set by this variation and in practice the optimum value of D is about 0.7 to 1.

The vector diagram shows the phase deviations and relationships around the phase detector, where $\theta_r(t)$ is derived from eqn. 8.5. It can be seen that the peak phase difference between the two inputs increases as ω_m is increased and is normally greater than the peak input phase modulation θ_i. For a $\pm 90°$ triangular modulation, the peak fundamental input is $8\times 90/\pi^2 = 70°$ and it can be shown that for the suggested range of D the worst case instantaneous difference is approximately 90°. Even so, this considerably exceeds the linear range of a 'sinusoidal' phase detector and since most suitable circuits give such a response, non-linearity effects must be allowed for. These are difficult to analyse exactly, but may be summarised by saying that the pull-out range is reduced by approximately 0.7 and the effective damping factor is increased above the value derived from a small-deviation analysis. The non-linearity of the phase detector also acts as a limiter on the phase deviation and therefore tends to reduce the peak of the triangular modulation, and this, in conjunction with the bandpass characteristic

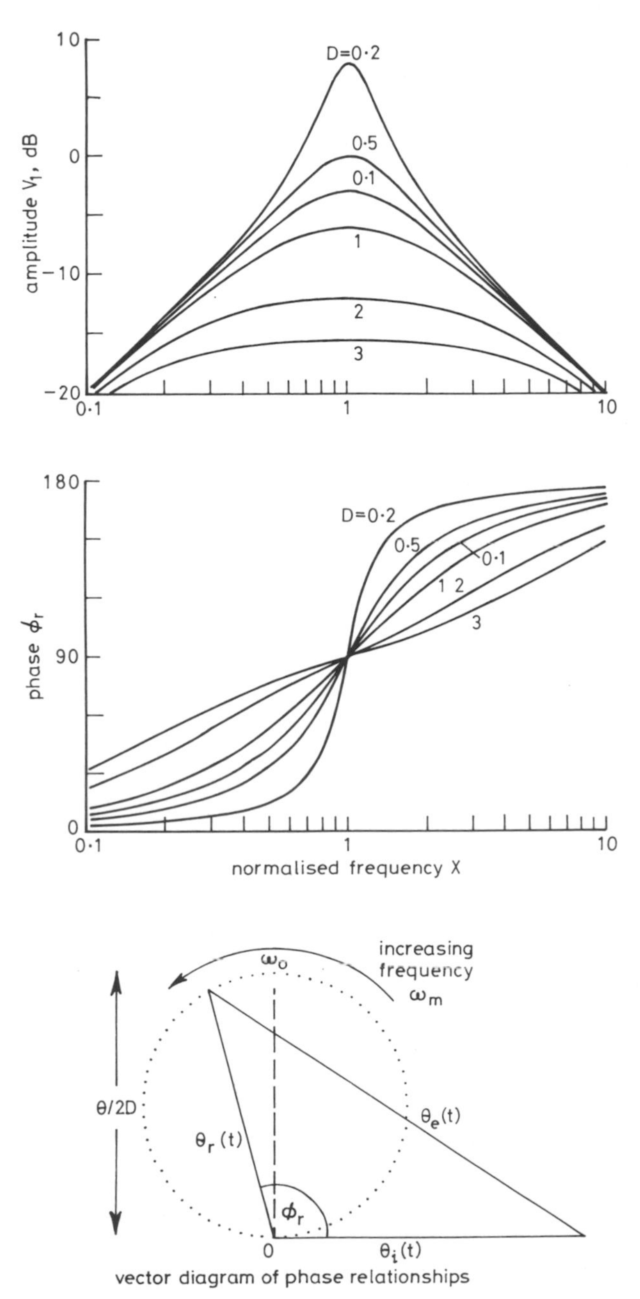

Fig. 8.4 *Amplitude and phase relationships of a second-order phase-locked loop with a phase-modulated input signal*

of the loop, results in an output to the second loop which is a fairly good sine wave as indicated in Fig. 8.2.

Any error in the mean frequency of the input signal will generate a d.c. output from the loop, and this may be displayed to the operator to indicate any Doppler shift or maladjustment of frequency standards. Alternatively, the voltage may be used to control the signal or reference frequencies in an automatic frequency control loop as discussed in Section 12.1.3.

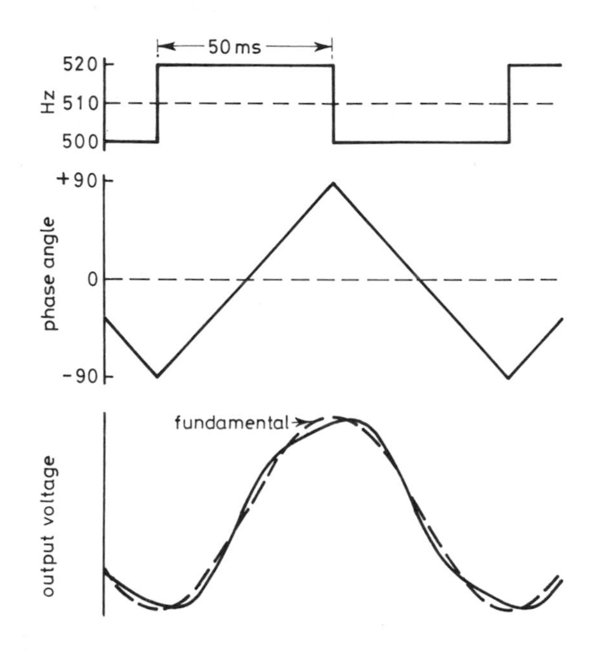

Fig. 8.2 *Synchronising signal and output of first phase-locked loop*

8.8 Second Phase-Locked Loop

The purpose of the second PLL is to pull the reference frequency waveform at ω_m into lock with the modulation waveform derived from the first loop. Once synchronisation has been achieved a series of checks is carried out to ensure that the signal being operated on is a correct standby signal of adequate amplitude, and that the synchronism is to an adequate accuracy. If all checks are satisfied then a phasing pulse is output from the synchronising system to the demodulator timing clock and pulls it into the correct phase. It is essential that the checking system is carefully designed to impose rigorous limitations on the conditions under which a phasing pulse may be output. In intervals between the reception of correct standby signals, the synchronising circuit is subject to the data stream (which may include elements at the same frequencies as are used for synchronising) and to high levels of noise or interference, particularly if the transmitted signal fails. It is essential that no spurious phasing pulse should be output under any of these conditions.

In the design for the Mark 6, these checks were based on measurements of the d.c. voltage from the phase detector in the second loop, and also from a second phase detector driven in quadrature with the first. The two outputs combined (preferably in the form of the RMS of the two voltages) give a direct measure of the amplitude of the 10 Hz signal received from the first loop. A control line from the matched filter system indicates whether either of the two synchronising tones is being received. A phasing pulse can only be output if the following criteria are satisfied.

(a) The amplitude of the 10 Hz waveform from the first loop exceeds a fixed threshold.
(b) The d.c. voltage from the loop phase detector is below a certain threshold (in conjunction with (a) this indicates the accuracy of synchronisation).
(c) Either of the two synchronising tones is being detected by the matched filters.
(d) The conditions (a), (b) and (c) have been met continuously for a certain minimum period.

The absolute levels for each of these criteria are not discussed here since they are not amenable to analysis and must be determined empirically, preferably with the aid of a channel simulator. Sufficient to say that the values finally used would seem to satisfy the requirements in that no spurious phasing pulse has been observed in any conditions, either in the field or the laboratory.

Check (a) serves a dual purpose. It has already been noted that the phase shift through the first loop varies with signal amplitude, so the imposition of a lower limit on the viable signal not only prevents spurious phasing pulses on low or false signals, but it also limits the extent to which timing variations in the first loop affect the timing of the phasing pulse. The overall performance achieved is that the system will pull into lock and output a phasing pulse in less than three seconds on a clean signal of adequate amplitude and within six seconds on virtually any signal which is viable for demodulation in the filters. The worst-case variation of the phasing pulse timing for all conditions of fading, noise etc., is less than about 6 ms total and in the vast majority of practical circumstances the timing error on a phasing pulse can be assumed to be less than ± 2 ms on a preset 'mean'.

8.9 References

1 MOSCHYTZ, G. S.: 'Miniaturised RC filters using phase-locked loops', *BSTJ*, May 1965, pp. 823–870

9

Practical Matched Filters

9.1 Lossless Resonant Circuit

In previous chapters the basic function of the matched filter has been described in terms of a resonant *LC* circuit, or single-pole bandpass filter, and although this approach may be considered mathematically and technologically obsolescent, the technique was employed in the early Piccolo and Kineplex equipments and may still be useful in some applications, thus justifying a brief description.

Although the description in Section 2.3 (Fig. 2.3) assumed zero losses, some tolerance is allowable on this condition since the circuit is initialised at the beginning of each element by the quench process. If losses are finite and positive, any oscillatory energy will (in the absence of an input signal) decay during an element period by a 'non-linearity factor':

$$\alpha \simeq \frac{T}{4CR} = \frac{\pi m}{2Q}$$

where

m = number of cycles of signal frequency in an element period T.
R = total effective parallel losses.
Q = effective circuit Q-factor ωCR.

If positive feedback is applied to cancel circuit losses (as will normally be required), R and Q are the total effective parameters including the effects of feedback, and if this is excessive the circuit will be oscillatory and R and Q are negative. A free oscillation will then increase by the same amount during an element period. The circuit should therefore be designed to maintain α within set limits (say $\pm$ 5%) over all circumstances. This will probably involve gain stabilisation of the active element and possibly temperature compensation. The circuits should be as linear as possible over a large amplitude range and no 'self-bias' components included. A simplified schematic of a transistor circuit embodying these principles is shown in Fig. 9.1, similar to that used in the early Piccolo equipments.

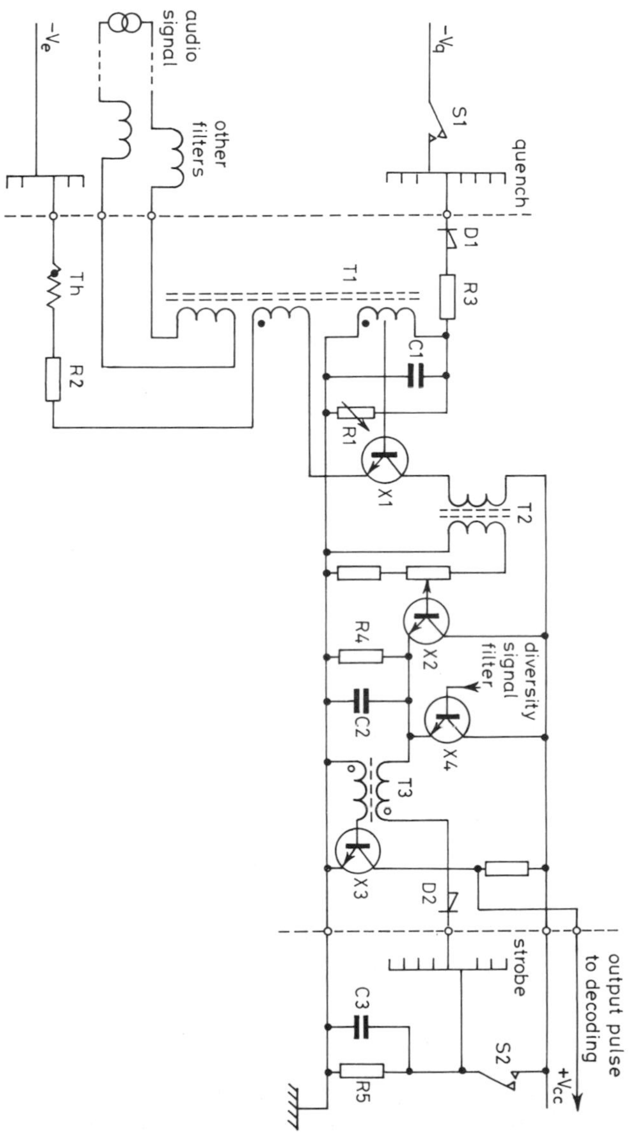

Fig. 9.1 *Matched filter based on resonant circuit (as used in early models of Piccolo)*
S1 and S2 are electronic switches controlled by receiving system timing circuits

The output envelope is detected by the second transistor and applied to one input of the comparator. At the end of each element the 'strobe' input (common to all matched filters) falls at a finite rate and the first filter to generate a comparator output transition indicates the selected frequency.

Immediately a filter has been selected, the strobe waveform is terminated and the 'quench' input (common to all filters) falls and reduces the energy in the filter circuit to zero. With careful design, the total duration of the strobe and quench periods may be reduced to less than 3% of an element.

There are a number of design problems with this circuit, mostly arising from the defects of transistors and diodes from their idealised concepts, and the tolerances on wound components, but a similar circuit in more modern technology would be appropriate in some applications.

Note that any frequency instability of the filter reactive components must be included in consideration of the effects of frequency errors (see Sections 3.5 and 10.4.3).

9.2 Quadrature Modulation

Consider the system shown schematically in Fig. 9.2*a*. The signal waveform $E(t)$ is input to two modulators, in each of which it is multiplied by a reference sine wave of amplitude A at a frequency ω_r (which is the required 'resonant' frequency of the filter), the two reference waveforms being 90° out of phase. Then

$$E_a = AE(t) \sin \omega_r t \quad (9.1a)$$

and

$$E_b = AE(t) \cos \omega_r t \quad (9.1b)$$

In the particular case where $E(t)$ is a sinusoidal waveform of frequency ω_s which is near to ω_r, it may be assumed that the subsequent integration removes all higher-frequency components of E_a, including the signal and reference waveforms, thus if

$$E(t) = E \sin(\omega_s t + \theta)$$

then

$$\begin{aligned} Ea &= E \sin(\omega_s t + \theta) A \sin \omega_r t \\ &= \frac{EA}{2}[\cos\{(\omega_s - \omega_r)t + \theta\} - \cos\{(\omega_s + \omega_r)t + \theta\}] \end{aligned}$$

If the integator output is assumed zero at $t = 0$, then at time T ($T \gg 1/\omega_s$):

$$V_a = \frac{EA}{2}\int_0^T \cos(\Delta\omega t + \theta)\,dt \quad \text{where } \Delta\omega = (\omega_s - \omega_r)$$

$$= \frac{EA}{2\Delta\omega}\{\sin(\Delta\omega t + \theta) - \sin\theta\}$$

Similarly

$$V_b = \frac{-EA}{2\Delta\omega}\{\cos(\Delta\omega t + \theta) - \cos\theta\}$$

The final output is given by

$$V = (V_a^2 + V_b^2)^{\frac{1}{2}}$$

$$= \left(\frac{EA}{\Delta\omega}\right)\left(\frac{1-\cos\Delta\omega T}{2}\right)^{\frac{1}{2}}$$

$$= \frac{EAT}{2}\ \frac{\sin(\Delta\omega T/2)}{\Delta\omega T/2} \tag{9.2}$$

Except for the gain factor, this equation is identical with eqn. 3.1.5, thus the system of Fig. 9.2*a* is fundamentally identical (within the limits of the analysis) with that of Fig. 2.3*a* with zero losses.

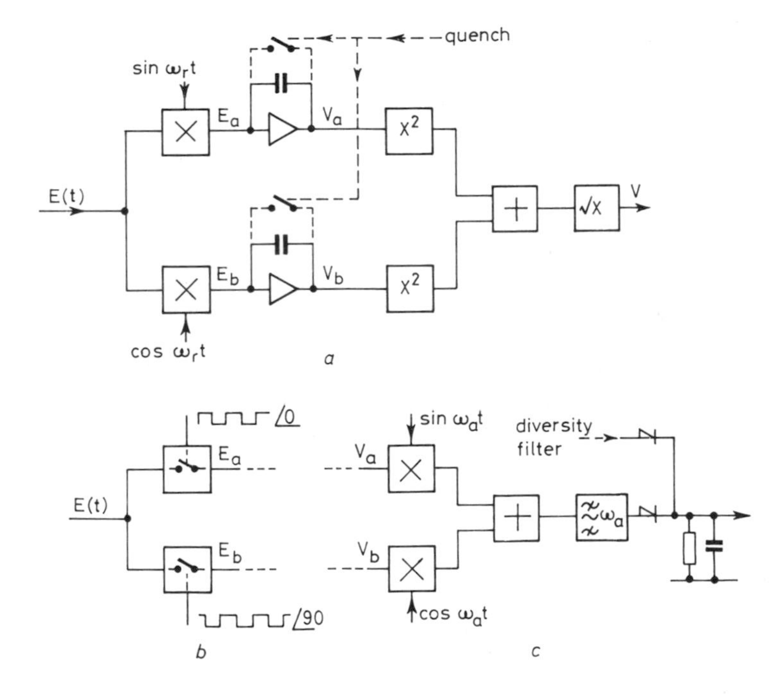

Fig. 9.2 *Basic principles of quadrature-modulation matched filters*
a Basic quadrature-modulation matched filter
b Switched modulators
c Root mean square extraction

9.3 Quadrature Switching

In a practical design, the system of Fig. 9.2*a* may be made considerably more economical if the two modulators (assumed in the above analysis to be four-quadrant multipliers) are replaced by simple switches (such as analogue FET switches, or, in some cases, binary gates) as shown in Fig. 9.2*b*. The reference frequency is now applied as a binary waveform controlling the switch. When the switch is closed $E_a = E(t)$, and when it is open $E_a = 0$, which is mathematically equivalent to multiplying the signal by unity and zero. The switching waveform can be analysed into a series of components at the frequency ω_r and its harmonics, therefore eqn. 9.1*a* is now replaced by:

$$E_a = E(t)\ (A_0 + A_1\sin\omega_r t + A_2\sin 2\omega_r t + A_3\sin 3\omega_r t + \ldots + B_1\cos\omega_r t + B_2\cos 2\omega_r t + B_3\cos 3\omega_r t \ldots) \quad (9.3)$$

If $E(t)$ is limited to a narrow bandwidth around the frequency ω_r and all higher-frequency products in V_a are removed by integration, the subsequent equations are as before, and the two systems are identical.

A special case requiring consideration occurs when bandwidth of the signal has been limited to comply with this requirement, but the signal has subsequently suffered distortion (by amplifier overload for instance). This will generate harmonics of the signal, which then interact with the corresponding harmonics in the reference waveform to produce spurious low-frequency outputs. This problem is considerably reduced if the switching is balanced to reject even harmonics, and if the switching waveforms contain no third harmonic. The latter may be achieved by closing the switches for one third of each cycle, rather than one half (see below).

9.4 RMS Extraction and Comparison

Although the functions of deriving the square and square root of a d.c. voltage may be carried out by analogue techniques, this can be relatively expensive, particularly if a wide dynamic range is required. An alternative approach is shown in Fig. 9.2*c*, in which the integrator outputs V_a and V_b modulate quadrature waveforms at an auxiliary frequency. The resulting waveforms are linearly added and then peak detected. It is easy to demonstrate that this output is proportional to the root of the sum of the squares. In principle, the frequency of the modulating waveforms is irrelevant, and it is sometimes convenient to use the same waveforms at the reference frequency as are used in the first pair of modulators. It is possible for the second modulators also to be in the form of analogue switches so that the waveform after addition consists of the sum of two square waves with a 90° phase difference. In this case it is essential that a bandpass filter is included before detection in order to eliminate d.c. and all harmonics of the waveform (although the requirements for this filter may be reduced by using switching waveforms with no third harmonic as described above).

The peak detectors may be of any form, the circuit constants being chosen to follow the envelope as faithfully as possible through a null at the end of an element (see Fig. 2.3) while keeping to a minimum the ripple from the first modulator and auxiliary frequencies.

The total overall accuracy of a set of matched filters is subject to three limitations:

The spread (total variation) of the gains of each individual path from audio signal source to detector output;
The ripple voltage on the peak detectors;
The accuracy of comparison.

The comparator circuit may be designed with appreciable hysteresis ('backlash'), the sole criterion of accuracy being the input offset at the instant of operation, i.e. the differential voltage between the strobe source and the detector output required to initiate the transition. A fixed offset (as generated by the diode D2 in Fig. 9.1, for instance) is of no consequence, but any difference between the offsets of the different matched filters will reduce the accuracy correspondingly. In this respect the circuit of Fig. 9.1 is not good, particularly at low levels, but modern operational-amplifier comparators allow this error to be reduced to negligible proportions.

9.5 Generation of Reference Waveforms

The effective 'resonant' frequency of the matched filter described is controlled entirely by the frequency of the reference waveforms applied to the first modulators. Since the maintenance of a high frequency stability is essential in MFSK systems, these waveforms should be generated by frequency synthesis from a higher reference frequency and the problem arises of generating a series of audio frequencies spaced at a constant interval (typically 10 or 20 Hz) to generate a series of frequencies given by:

$$f_r = f_c + nf_i$$

where f_c is the lowest frequency required (the 'channel frequency'), f_i is the interval between adjacent frequencies, and $n = 0, 1, 2 \ldots$ up to the number of tones required.

Four waveforms in quadrature are required at each frequency, and these waveforms should contain no third harmonic. A method of achieving this in a reasonably economical manner is shown in Fig. 9.3*a*. The crystal oscillator output (divided if necessary) is first divided by two, generating a 50:50 square wave, the positive transitions (say) of which are used to generate the channel frequency, and the negative transitions the frequency increments. This ensures that no transition generated in the one stream will coincide in time with one generated in the other stream. The first stream is divided as required to produce a waveform at mf_c where 'm' is a large integer (discussed below). From this waveform is generated a narrow positive pulse (shorter in duration than a half-period of the oscillator source). The second pulse stream (using the other transition of the first divider) is divided by such a number as to generate a waveform at $8mf_i$ (for $M = 16$ or less). This is then successively divided by two to produce a series of waveforms at binary intervals of frequency down to mf_i. In each case the positive transition of a divider is used to trigger the next divider stage, and the negative transition generates a narrow pulse as before.

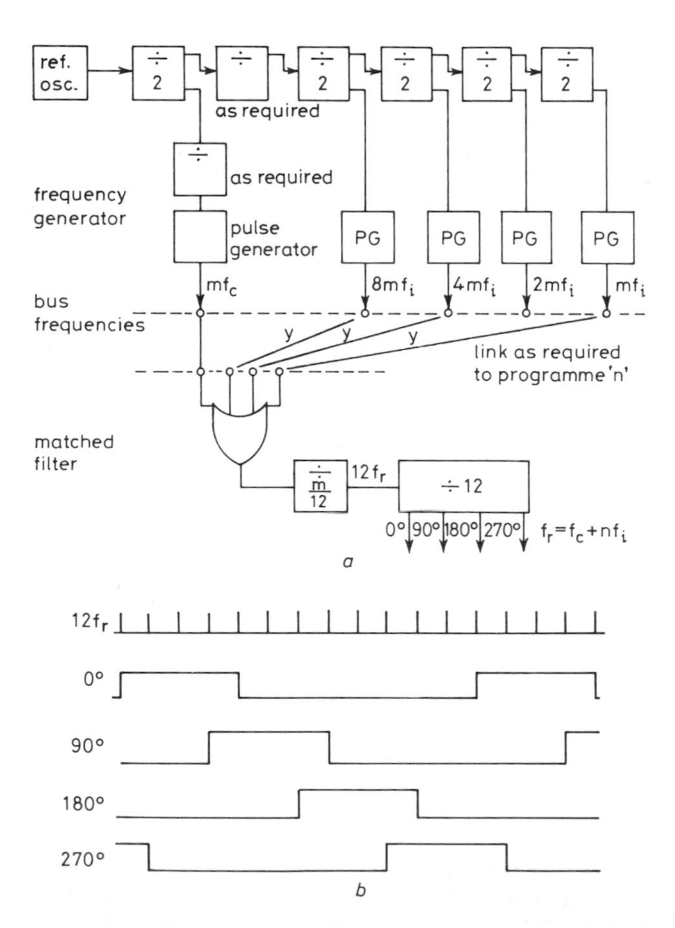

Fig. 9.3 *Synthesis of reference waveforms for matched filter*
a Method of frequency synthesis
b Output waveforms

The outputs from the pulse generator circuits drive a series of frequency 'bus' wires to the matched filter circuits. Note that each bus wire carries a stream of narrow positive pulses at the particular frequency required, and that no two pulses on any of the bus wires overlap. To derive a single pulse stream at a required frequency it is only necessary to combine the required pulse streams in a multi-input OR-gate and then to divide the resultant pulse stream by m.*

The value decided on for m is a compromise. The composite pulse stream from the combining gate is at the correct mean pulse repetition frequency but will have a pronounced timing jitter which is successively reduced as the frequency is divided. Too small a value of m leaves too much jitter on the final reference waveform, leading to spurious responses of the matched filter. Too high a value of m means that the initial oscillator frequency must be very high and the bus pulses very narrow, leading to higher power consumption and dissipation in the electronic circuits and greater difficulty in distribution of the waveforms. The value of m should be a multiple of 12, and measurements on an experimental system suggest a minimum value of about 80 for an acceptable level of spurious response. Practical figures therefore lie between $m = 96$ and about 360. The required division is carried out in two stages, firstly by $m/12$, and then in a 'divide by 12' circuit which generates the four waveforms as shown in Fig. 9.3*b*. These are at the required reference frequency, at 90° phase intervals and in each case the duration of the 1 state is one third of the period.

9.6 Amplitude Limiting

The use of the technique described in the preceding paragraphs enables a series of matched filters to be realised in an economic manner using standard integrated circuit components. A very high frequency stability and a dynamic range of better than 30 dB can be obtained without the use of precision active components, presets or nulling adjustments.

The dynamic range may be further extended without difficulty by taking advantage of the third harmonic rejection technique described above, in conjunction with controlled peak limiting of the input signal at a specific level as discussed in Section 5.4. It has also been found convenient to limit the detected output from each matched filter at about the level produced by an on-frequency audio signal just at the onset of input limiting.

9.7 Complete Matched Filter

The input audio amplifier is symmetrically peak limited in a controlled manner (using Zener diodes) and drives a complete bank of matched filters in parallel, an

---

*The same basic technique may be used to synthesise the MFSK tone output stream in the sending modem, the bus pulse streams being selected dynamically by logic gates operated by the coded data.

inverted signal also being generated. The original signal is modulated in an FET switch as described in Section 9.3 and Fig. 9.2*b* by the reference signal from Fig. 9.3*b* at zero phase, and the inverse with the reference at 180°. The outputs are combined into a single integrator. The result is a balanced modulator with no sensitivity to d.c., even harmonics or the third harmonic. The same two audio signals are similarly modulated with the 90° and 270° references and the sum integrated to extract the quadrature component. Each integrator capacitor has an analogue FET switch across it, which is closed for a brief period at the end of each element to quench the circuit to zero. The bipolar output voltages then modulate quadrature binary waveforms at an auxiliary frequency of approximately 5 kHz (to reduce detector ripple and give simple filtering). The resultant square waves are added, filtered to remove d.c. and all harmonics, then peak detected. The output from the peak detector forms the output of the filter as described in Section 3.1, eqn. 3.6 and Fig. 2.5.

Selection diversity (see Section 7.2) is achieved by providing a separate audio amplifier chain feeding an identical bank of filters. The two filter circuits at the same frequency use the same reference waveforms and the two peak detectors feed into a common detector load which therefore follows the higher of the diversity signals at that tone-frequency. The detector output is applied to one input of an operational-amplifier comparator. The complete circuit uses only standard linear and logic integrated circuits and FET analogue switches. For miniaturised equipment or where large quantities are required, the circuit is well suited for production in the form of one or two encapsulated hybrid modules.

The overall dynamic range extends from more than +10 dB on the input limiting level to less than −30 dB on that level, and, using fixed passive components of ±1% tolerance, the tolerance on the overall gain of each filter is typically better than about ±0.5 dB. The total time required to strobe and quench the circuit is less than 3% of an element duration (1.5 ms in the case under discussion) and could be reduced still further if that were proved necessary. Spurious effects such as unbalance between quadrature paths and responses to jitter components on the reference frequencies are typically less than 2%.

9.8 Digital Signal Processing

With the tremendous advances made in recent years in digital signal processing, the question is frequently asked whether an MFSK system could be realised by such techniques. In the Piccolo Mark 6 all the send circuits are digital up to the final analogue filter (which could have been realised digitally if convenient). In the receiving system all frequency synthesis and clock waveform manipulation is digital, the only analogue circuits being the audio amplifiers, synchronising system and matched filters. These functions could all, in principle, be realised digitally (in the sense of converting the audio signal into digital form before processing), if this could be shown to be of any advantage. In considering this question, the first major factor to be established is that, for the range of parameters discussed in this book, there is no point at which the limitations of the technology used are such as to cause a deterioration in performance. More specifically, all relevant frequencies are derived by digital synthesis from a high-stability reference crystal oscillator, the dynamic range achieved using analogue techniques would be difficult to improve [1], the accuracy of synchronising is better than the timing variation which may be introduced by multipath reception, and so on. In short, the use of a different circuit technology is not likely to improve signalling accuracy under any practical circumstances.

Advantages in cost are difficult to analyse since the existing Piccolo units are fully developed and the circuit technology proven over a number of years, whereas an all-digital system is a hypothetical situation which cannot be accurately costed. The reliability of the Piccolo units should be high enough not to add appreciably to total system down-time, and equipment size is not a problem. Another consideration is convenience (in concept, maintenance etc.). In this book the modem is seen as an interface unit accepting an analogue audio signal from a radio receiver and outputting one or more low-speed telegraph signals to a corresponding number of conventional teleprinters or the equivalent. Within the context of that system there would seem to be little point in introducing a different technology into the modem unless it was justified on grounds other than the mere urge to keep in the fashion.

In the author's opinion there are two cases where the realisation of an MFSK receiving system in purely digital terms may be justified. The first is when the modem forms a part of a very much larger system (message storage, data processing etc.) realised entirely or mainly in digital techniques. The second is if a communication system requires a large number of filters, so that the cost and/or bulk of analogue processing makes it less attractive. Some examples of such systems are discussed in Sections 12.5, 12.6 and 12.7. For such a design a number of different techniques would need to be considered, including the possibility of serial processing in a high-speed time-multiplexed circuit rather than parallel processing in a number of separate matched filters. Three or four alternatives may be suggested for investigation such as:

Digital realisation of the quadrature-modulation techniques described above.
Digital cross-correlation or fast Fourier transform techniques.
Digital filtering.
Quasi-analogue filtering using charge-coupled devices (such as 'bucket-brigade' delay circuits).

9.9 References

1 FELLGETT, P. B.: 'Some comparisons of digital and analogue sound recording', *The Radio and Electronic Engineer*, Feb 1983, **53**, No. 2, pp. 55–62.

10

Data Control and Frequency Tolerances

10.1 Data Source Control

This chapter will discuss one or two points concerning the data stream before the modulation process in the send equipment and after the demodulation process in the receive equipment, and also the allowable tolerances on various frequency standards. Many of these problems are not peculiar to MFSK systems or to h.f. telegraphy but are included here as possibly providing a useful reference to appropriate techniques.

A general problem applying to all synchronous communication is that of the control of input data to the fixed data rate of the communication link itself. Where the data source is an integral part of the communication equipment this is usually not a serious problem, but in many cases of h.f. telegraphy the source is remote, or is a mechanical device such as a conventional teleprinter or tape reader, in which the data rates may be relatively inaccurate (see Section 2.7). For present purposes any data source may be classified under one of three headings:

(a) *Uncontrolled and manual control.* The source outputs data into the sending system at a rate determined by its own internal clocks or mechanism and there is no method by which the modem equipment can vary this, but there may be a switch by means of which an operator can stop the data stream.
(b) *State-controlled or clutch controlled.* Many mechanical tape readers have a solenoid-operated clutch which disengages the tape-reading and tape-advance mechanisms from the driving motor. The data source may therefore be stopped and started by a control voltage from the modem equipment. In some cases the control may be relatively sluggish and the response may be delayed by one or more character periods. Electronic data sources may provide similar facilities

although in this case an appreciable delay is unlikely. While the data source is active the character rate is uncontrolled.
(c) *Pulse controlled.* Most modern electronic data sources provide alternative methods of data rate control, including a pulse-release mode in which a single input pulse on a control line will release a single data character or byte. The time delay between the control pulse and the receipt of the corresponding character may vary.

Note that in nearly all cases the control of the data source is carried out on a character basis, and the modulation rate (element length) of the source is derived from its own internal clock. However, normal commercial tolerances are such that variations in element length about the specified figure are easily allowed for in the input system of the modem.

In order to cater for the first two modes of operation, it is essential that the input circuits of a sending modem should operate in an asynchronous (start–stop) mode in which each data character is read individually into a store by a sequence of pulses initiated by the leading edge of the start element. The output from this first storage is conveniently taken in a parallel mode with one line for each data bit (there is no need to store the start and stop elements) and for the rest of the system the data may be assumed to be handled in parallel characters or bytes. The parity element in the ITA-5 code (or any similar binary check) is ignored for MFSK coding purposes, for reasons discussed in Section 11.3.

If the equipment is required to operate with relatively unstable sources, and particularly if it needs to handle cryptographic or unformatted telemetry information (see below) further buffer storage is necessary, and a suitable system is shown schematically in Fig. 10.1. For most applications a store capacity of 32 bytes is adequate and this is assumed in the following discussion. The store operates on a 'first in, first out' (FIFO) basis in which the first character input ripples through the store almost instantaneously to the output stage. Subsequent

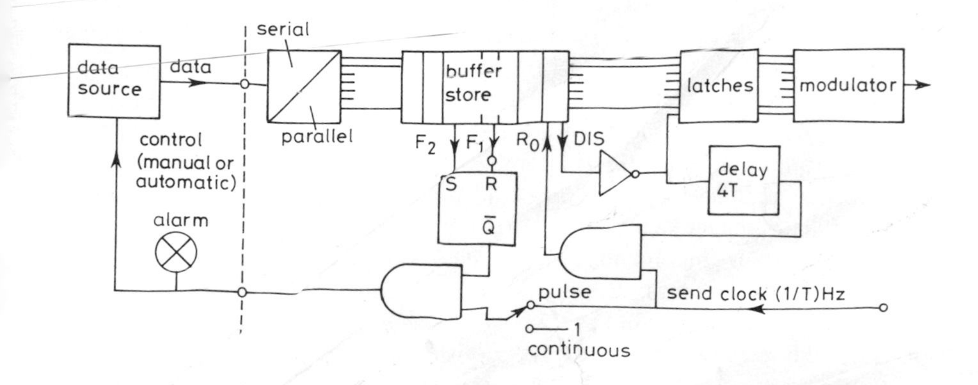

Fig. 10.1 *Block schematic of a system of data source control*

input characters similarly ripple through to occupy the lowest empty stages. The application of a read-out (RO) pulse transfers the character from the output stage to the output latches where the data is then available on the output bus lines to the modulator. This transfer empties the final stage and all the contents of the store move down one stage. The presence of a character in the final stage of the store is indicated by a data-in-store (DIS) state. The timing of the readout pulse is controlled by a 'send clock' pulse stream at the required character rate of the communication channel, assumed 10 Hz in the discussion, so that under normal conditions with the store partially full, the data is read out of the store at a fixed rate of 10 characters per second.

If the data source is uncontrolled, and is feeding in at a mean rate greater than 10 characters per second, then the store will slowly fill up, and, if the process is continued until it overflows, data will be either lost or corrupted (depending upon the design of the store). To prevent this, two 'flag' indications are provided on the store, typically at about one-third and two-thirds of its contents (although these levels are by no means critical and the Piccolo Mark 6 design uses flags at 8 and 16 characters, respectively). If the contents of the store exceed the upper flag an 'excess data' alarm state is set and is not cancelled until the contents of the store drop below the lower flag. This indication can be used in several ways, depending upon the facilities available. In the simplest case of a source which may be started and stopped manually, the indication can be used to operate a lamp, bell or buzzer, which notifies the operator that the source should be stopped, and when the indication is cancelled the source is restarted. Where the source may be state controlled this may be done automatically. Under such circumstances the store will fill slowly (at a rate depending upon the difference between the source rate and the output rate) up to the higher flag, then empty rapidly at the output rate down to the lower flag. Thus with either manual or state control the store will not overfill if the source is running fast.

Where the data source may be pulse controlled a flag system may not be required, the data being 'called-out' continuously from the source by a 10 Hz pulse synchronous with the RO clock. However, it is advisable to disable the call-out pulse with the excess data state since this gives absolute control under all circumstances, such as the introduction of an error code (see Chapter 11) which reduces the output data rate.

If at the instant that the RO pulse is applied there is no data character in the store, the 'standby' output line is energised for the following character period; the number of parallel data lines into the modulator being one more than the number of data elements in the input byte (excluding start, stop and parity elements). Thus for uncontrolled, manual or state-controlled sources running slower than the system data rate, one or more standby characters will be inserted at intervals in the data stream.

A particular case which should be considered is when the data source is operating at the correct mean rate but with some timing jitter. If there is nominally a single character in the store, a coincidental relationship between the timing of the RO pulse and the time of arrival of characters may cause an occasional character to be read out immediately it is received, or with almost 100 ms delay, causing the store contents to jitter down to zero or up to two. Standby characters may then be inserted erratically. This does not matter for most normal telegraphy traffic, but there may be cases (see below) where it is inconvenient to have standby characters inserted during a message. This effect can be eliminated by introducing a delay in the read-out as shown in Fig. 10.1. When the store is empty (as indicated by the absence of the DIS condition) the RO pulse is disabled, and when the first character is indicated by the presence of the DIS condition, this indication is delayed by a number of character intervals (say four) before the RO pulse is enabled. If the character is the first of a continuous data stream, this delay allows four characters to accumulate in the store before the first one is read out, and minor timing variations can then only cause a jitter of store contents between three and five and no standby characters are inserted. If the data source is running slow, the data stream is maintained until four characters have been 'lost', when a block of four standby characters is inserted before the stream begins again. With completely uncontrolled sources, and if the insertion of a standby character is as undesirable as a store overflow, this delay may be made equal to half the store capacity, providing an equal protection for both fast and slow data sources. A minor disadvantage of the delay is that it introduces a corresponding real-time delay in the communication channel which may be inconvenient for hand-keyed Simplex or ARQ operation (see Section 11.2). An advantage is discussed in Section 10.3.1.

10.2 Standby and non-valid characters

It was suggested in Chapter 8 that the synchronising information should be conveyed by the standby signal and that this should take the form of one or more additional characters to the basic input alphabet. For the simpler telegraphy systems the receipt of a standby character causes the output data line from the receiving modem to be held in the rest condition (holding mark) for a full character period, in effect omitting a character. For repeated standby characters the printer remains at rest. Exceptions to this are discussed below.

Section 6.6 discussed the existence of non-valid characters (NVC) which cannot be generated by the sending circuits, but which may be received by the receiving circuits if one of the tone elements is misinterpreted. The receive decoding system must include provision for recognising and treating such errors. In most cases the choice lies between omitting the character from the output stream (as for the standby character), outputting a character which is ignored by the printer (such as DEL in ITA-5 or character 32 in ITA-2), or outputting a specific 'error indicating' symbol. The choice is not important for normal telegraphy, but may be of importance in cryptographic or telemetry operation.

10.3 Cryptographic and Telemetry Data

The above design is adequate for normal plain-language telegraphy but there are applications which may require special facilities, notably those involving cryptographically protected or unformatted telemetry data. In either case the input may consist of a continuous stream of quasi-random characters without format or interruption, the significance of the data being extracted at the receiving end by either a decrypting 'key stream' or a stored telemetry format. In either of these cases the deletion or insertion of a character can cause a major error of interpretation and a temporary or permanent breakdown of the interpreting system. In these circumstances care must be taken to study the implications of the use of standby and non-valid characters. Two possible modes of treatment are suggested, according to the type of data control available at the sending end (and in general-purpose equipment, both modes should be available).

10.3.1 Normal Operation

Ideally, a link which requires such special considerations should operate from a pulse-released source so that no standby characters are inserted into a continuous data stream. Then every standby or NVC character received can be assumed in the first instance to have been caused by a detection error and a 'fill' character of any kind (typically DEL in ITA-5 or character 32 in ITA-2) output to line so that no slip is caused. If four such characters are received in succession, it is recognised that this must be the end of a data stream and all successive standby characters are deleted. Any data character received resets the count to zero.

This mode may also be used in conjunction with a state controlled or manually controlled data source where this is of relatively high stability. In such cases, if the source is running slow, a character slip will occur each time that the store completely empties and one or more standby characters is transmitted. The length of the data stream which elapses before this occurs may be considerably extended by the use of a time delay in the store read-out as described above. If the data input rate is low by a factor r, and there is a delay of n characters in the input store read out, then the minimum possible duration of an uninterrupted input data stream is $(n\text{-}1)/r$ characters. For example, if the source is 0.1% slow, and the delay is 4 characters, then at least 3000 characters may be sent (starting from an empty store) without being interrupted by a standby character.

10.3.2 Slow Source Operation

If the data is from an uncontrolled or state controlled source which is of relatively poor stability the character slips caused by the store becoming empty may occur at an unacceptably high rate. The 'slow source' mode of operation causes all standby characters received to be deleted from the output data stream. Now no character slip will occur when standby characters are transmitted but a slip will occur each time that standby is received in error for a data character (or vice versa). The time interval between such events (and therefore the duration of character stream which may be sent without slip) will be random, the mean being dependent upon the mean error rate of the communication link. The mean interval can be roughly calculated by assuming that all errors are equally probable, so that if the mean character error rate is p, the probability of receiving standby as an error for a data character is approximately p/A_e (see eqn. 2.3) and the mean time between slips the inverse, A_e/p characters. Comparison of this figure with that derived above for the normal system will indicate the optimum mode of operation for given conditions.

10.4 Frequency Tolerances

The question of tolerances on tone frequencies and data rates has recurred throughout this book and some confusion could well arise. This section is intended to summarise and clarify the situation. Fig. 10.2 indicates that in a typical h.f. telegraphy link using SSB modulation as many as five frequency standards or sources may be involved, although in any particular case the number is likely to be less since the same standard can be used for several purposes:

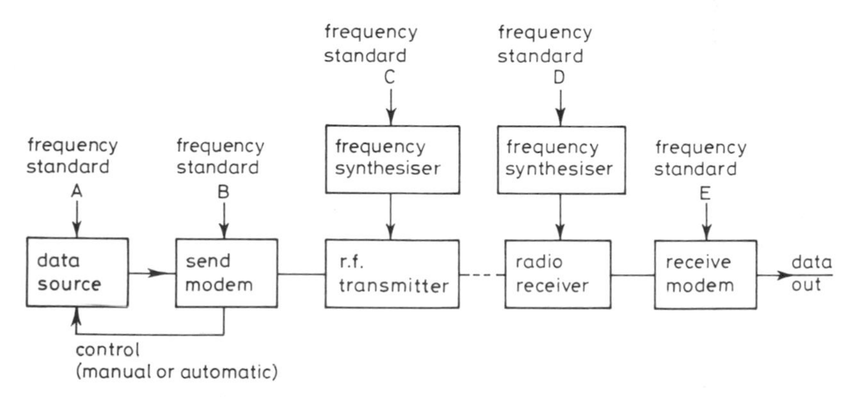

Fig. 10.2 *Block schematic of basic radio telegraphy link showing frequency standards*

A The mean data rate of the data source may be controlled by a crystal oscillator (in a modern equipment), the rate of rotation of an electric motor in the case of a conventional tape reader, or the operator's typing rate in direct keyboard operation.

B The frequency standard of the sending modem will determine the accuracy of the sending data rate. It will also determine the accuracy of the audio tone frequencies generated but this effect is normally negligible compared with standard C.

C The frequency standard controlling the transmitter synthesiser governs the accuracy of the transmitted radio frequency carrier to which the tone frequencies are added (assuming USB modulation). For analysis purposes it is assumed that all frequency errors are directly proportional to the carrier frequency.

D The frequency standard to the radio receiver governs the accuracy of the radio frequency system so that the accuracy of the received audio frequencies output from the receiver is proportional to (C – D).
E The frequency standard to the receiving modem governs the accuracy of the element-rate clocks controlling the matched filter detection process. Where quadrature-modulation detectors are used (see Section 9.2) it will also govern the accuracy of 'tuning' of the matched filters, but again this effect should normally be negligible compared with the effect of standard D on the accuracy of the signal frequencies (but see Section 9.1).

There are therefore three different frequency errors to be considered.

10.4.1 Radio frequency

In Section 3.5 it was shown that a frequency error equal to 10% of a tone interval will cause a deterioration roughly equivalent to 1 dB of signal-to-noise ratio. Taking this as the maximum permissible, the equivalent tolerance on frequency standards C and D is given by:

$$a = \frac{0.1}{Tf_c} \tag{10.1}$$

where T is the duration of a matched filter element, f_c is the nominal carrier frequency, and a is the tolerance on difference in frequency between C and D.

Thus for an element length of 50 ms, and the highest h.f. frequency of 30 MHz, the sum of the tolerances on standards C and D would be:

$$a = \frac{0.1}{0.05 \times 30 \times 10^6} = 7 \times 10^{-8}$$

This is by no means an unreasonable tolerance for point-to-point operation of static equipment where the crystal oscillators in the frequency standards may be left continuously operational, and periodic checks made between the equipment at each end of a link. The situation is a little more difficult for tactical or emergency links which may be required to operate at very short notice, and possibly with severe power consumption restrictions. This application may require the use of high quality temperature-compensated un-ovened crystals. The difficulty is somewhat mitigated by the fact that most of such links operate on relatively short ranges where the frequencies used tend to be correspondingly low, and of course the system breakdown is not sudden at the suggested limit. For an alternative approach, see Section 12.1.3.

10.4.2 Element rate

The element rate of the sending system is determined by frequency standard B and that of the receiving system by standard E. Assuming perfect initial synchronism by the one-shot system described in Section 8.5, then once a continuous data stream begins, the matched filter detection system will begin to drift out of synchronism at a rate determined by the difference between the two. Section 3.6 suggests a deterioration of about 1 dB for an error of about 5%, while 3.7 indicates that this is pessimistic and in fact in many cases an error of at least 10% is permissible. Taking 10% as a tentative figure, the permissible tolerance is then given by:

$$b = 0.1T/T_s \tag{10.2}$$

where b is the maximum permissible tolerance on E (assuming B is accurate), T_s is the maximum duration of continuous data stream allowable, and T is the duration of element period.

Thus, for $T = 50$ ms and $b = 10^{-6}$, a continuous stream of at least 14 hours is permissible.

Again these figures may be considered to be pessimistic in that exceeding this time does not cause a sudden breakdown to the system, but rather a very slow deterioration in immunity to noise.

10.4.3 Data source

The frequency differential which must be accepted by the source control system is (B – A). The methods of control of the data source have already been described and it is pointed out that if the data source is running fast, then it is essential to provide some means of control to prevent the input store from being overfilled. The length of time before this will occur is set by the capacity of the store (less any reserve which may be introduced by the read-out delay). For most purposes this is adequate even for manual control but a particular case arises when an error coding system is included (see Chapter 11), when the data rate of the communication link may be reduced by 10% or more. Under such circumstances it is essential that some form of automatic control be used.

For normal plain-language text there is no reasonable limitation on the lower limit of the source data rate (including hand-speed keyboard operation). The restrictions for cryptographic operation have been discussed in Section 10.3.

11

Error Detection and Correction Techniques

11.1 Introduction

The introductory chapter of this book indicates that it is intended to strike a balance between the theoretical and the practical. Many experts in the field of error coding* will detect in this chapter a marked lack of the former. Prior to the work described here the author had no previous experience or tuition in this field and did not aspire to the mathematical ability to acquire such knowledge to the degree of expertise required, particularly since the published work on multi-level coding is infinitesimal compared with that on coding for a binary symmetric channel, and necessarily of a higher order of complexity. Furthermore, error coding for the Piccolo telegraph modems was always envisaged as an optional extra to cover special requirements rather than a fundamental factor in the communication system. The approach here is therefore blatantly pragmatic, and no attempt is made (other than in the most general terms) to compare the results achieved and the methods used with those already existing in the field of binary communication. None the less, it is believed that some of the concepts may be novel and the method of multi-level coding, either for multi-level or binary systems, may prove to justify further work in this field by those better qualified to discover its full potential.

This complex and specialised subject is therefore only considered here to a depth necessary to indicate how such techniques may be applied to MFSK links and to suggest lines of approach which may result in the design of effective coding for such links. For this reason some techniques such as soft-decision coding, convolution codes and concatenated codes are ignored, and discussion is

*Error coding: see 'Glossary of Terms and Abbreviations'.

limited to the concept of simple block codes. This does not imply in any way that more sophisticated techniques are not applicable to MFSK, but simply that insufficient work has been done in these areas to justify discussion, and extension from the simpler concepts of block codes is in most cases self-evident. The exclusion of soft-decision coding means that, for present purposes, error coding is carried out completely independently of the demodulation process itself (but see Section 12.1.5).

This chapter will first examine the present situation with reference to coding for binary channels, then examine how far this work is directly applicable to MFSK, then develop an approach to multi-level coding.

The basic concept underlying error coding is that additional information is transmitted in the data stream, and related to the data symbols in such a way as to form a check on their accuracy. The amount of additional information is normally expressed in terms of the 'redundancy' of the data stream, so that if the transmitted signal contains one element of check information for every nine elements of data it is said to be 10% redundant or to be a '0.9 rate' code. In 'block' coding a sequence of data elements is coded to produce the required number of check elements and a code block consisting of data plus check information may then be decoded as an entity independent of the rest of the message (although there is no necessity for the elements of a block to be sent in direct succession, see 'interleaving' below). The main repercussion of such coding on the design of a modem system itself is therefore the requirement for additions to the input data. This may result in a larger alphabet (see Section 2.5), a larger bandwidth or a reduction in output data rate.

From about 1950 to 1970 the subject of error coding was a playground for mathematicians and in the author's opinion it is unfortunate that much of this work was based on unrealisitc fundamental assumptions which often yielded misleading results. For instance, most of the work of that period totally ignored the fact that if the nominal output data rate from the system is maintained constant (the only grounds on which comparison may fairly be made), then the redundancy inherent in the use of an error code requires an increase in modulation rate and so inevitably increases the raw element error rate input to the decoding system. This effect can in very many cases nullify any advantage to be gained from the coding itself [1]. The fundamental mistake was to compare the error rates at the input and the output of the decoding process, whereas a more correct picture would be obtained by comparing the output from the decoder with the output of a non-coded system operating at the same *output* data rate, assuming the same signal-to-noise ratio and one of the standard demodulator error-rate relationships in Table 7.1.

Similar errors were committed in the field of practical measurements and comparison. For instance, one commercial FEC system operates in a synchronous mode with no start or stop elements in the data. This gives the system two major advantages, first in the elimination of signal power wastage in synchronising elements, and secondly in the elimination of the errors inherent in start–stop operation (see Section 2.7). In the published work describing this equipment, the results were quoted of comparison trials between the error coding system and a start–stop $7\frac{1}{2}$ unit link and the improvement obtained was attributed to the use of error coding. More detailed analysis of the figures [1] shows that the improvement attributable to these two factors would account for a large proportion of that claimed and that in fact the error coding system was considerably less effective than was suggested by the measurements (and these themselves showed a discrepancy of several orders of magnitude from the results predicted by theory).

More careful analysis in which such discrepancies are eliminated suggests that the margin between the calculated and measured performances of an error coding system is often sufficient to suggest that the use of such a system may be of doubtful benefit. The blunt fact is that a good error control system (whether FEC or ARQ, see below) can produce almost error-free copy from a signal of mediocre quality, but may result in a complete breakdown of communication on a poorer signal which would, without error coding, have still given acceptable accuracy. There is therefore a firm basis for arguing that for tactical and emergency links (where in effect 'poor communication is better than none at all'), a system without error coding is often to be preferred and that the concept of error coding techniques as a general-purpose means of compensating for a weak system is a doubtful philosophy.

Once such factors began to be appreciated, the pendulum swung the other way, and a number of papers attempted to compare the improvement to be gained from error coding with that from dual diversity reception, or use of different modulation techniques etc., apparently implicitly assuming that these choices were alternatives rather than independent factors. Although there are certain limited cases where this may be true (see, for instance, the Appendix) these are exceptional. Such an approach ignored the fact that there are three links in a typical communication chain, the radio system, the modem, and the error coding system, and provided that the characteristics of each system 'match' in major parameters (e.g. the radio receiver bandwidth is correct for the modem requirements, the modem will handle the required data rate etc.) each may be relatively freely selected from a number of alternatives. To attempt to equate, in performance terms alone, a change in one part of a system with a change of a totally different nature in another part (where the impact of capital costs, operational flexibility, compatibility with other systems and other imponderables may be totally different) can be extremely misleading.

11.2 Basic Concepts

There are three different ways in which redundant information may be used to improve the accuracy or reliability of the received information:

(a) *EI (error indication or error detection).* Typically, a code with about 10% redundancy can be used to detect all cases of a single element error in a block and some cases of multiple errors, but some cases of multiple errors will be undetectable and the probability of undetected output errors is finite. The detection of a block containing one or more errors (referred to as an 'error block') is indicated to the operator or information processing system to take appropriate action according to the facilities available.

(b) *ARQ (automatic request for repetition).* If suitable facilities exist (i.e. the link operates in a duplex mode, data storage facilities exist in both send and receive equipments, the sending data rates may be automatically controlled etc.) the error indications from an EI system may be used to initiate automatically a request for repetition of the corrupt data [3]. The redundancy under good signal conditions is of course no more than that required for error indication, but when errors occur the output data rate is decreased by the control signals and repetitions and in extreme cases may be reduced to zero. The undetected error rate is normally that of the EI code itself.

(c) *FEC (forward error correction).* By the use of a larger amount of redundancy (typically 30–50%) it is possible to correct single errors in a block, and sometimes a limited number of double errors (depending on the length of the code block). Some FEC codes also include the possibility of indicating the existence of larger numbers of errors, without necessarily being able to correct them.

Ignoring second-order effects and minor exceptions it is convenient to define the performance of a binary FEC code in terms of a mathematical model involving three parameters (n, k, e) defined as follows:

If, in a block of n binary signalling elements, a total of e errors or less is received, a byte of k bits of information is interpreted correctly. If there are $(e+1)$ or more errors the byte is incorrect.

Limiting relationships between these three parameters can be obtained and it can be shown for instance that a (10, 5, 2) code is not possible [2]. Known codes include the Hamming (10, 5, 1), the BCH (15, 7, 2) and the Golay (24, 12, 3) codes which will indicate realistic orders of magnitude. Performance curves for such a code are simple to derive, since the probability of a block containing uncorrectable errors is the probability of the occurrence of $(e+1)$ or more errors in n elements. If the element error rate on the input is p, the undetected byte error rate on the output is given by:

$$P_e = \sum_{r=e+1}^{n} C^n_r \, p^r(1-p)^{n-r} \tag{11.1}$$

Such a curve may be combined with the performance of the demodulation system as given in Table 7.1 to predict the overall performance of a modem and coding system with a given received signal-to-noise ratio. This has been carried out in Fig. 11.1, which shows three curves:

(a) The output character error rate for an ideal non-coherent FSK binary signalling system (system E in Table 7.1) carrying the ITA-2 telegraph alphabet in a 5-bit synchronous format at 100 Bd (data rate 20 characters/s).
(b) The same system but operating at 50 Bd (10 characters/s).
(c) The 100 baud system (a), with the addition of a half-rate (10, 5, 1) FEC code, giving an output data rate of 10 characters/s.

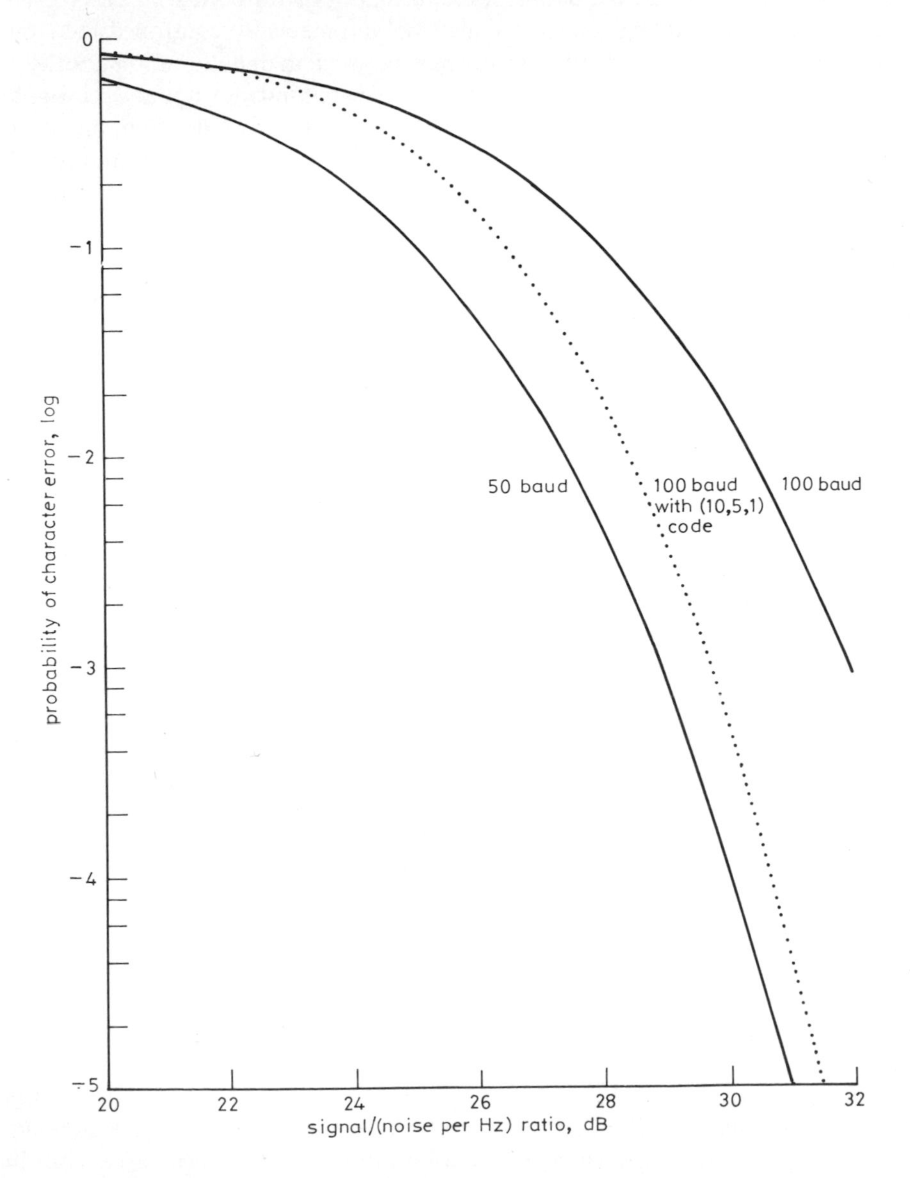

Fig. 11.1 *Character error rates of FSK systems with and without FEC coding*

If the error rate of system (a) is not acceptable, then either of the alternatives (b) or (c) is possible and will halve the data rate and give a reduction in errors. The improvement derived from the FEC code is given by eqn. 11.1, while that from the slow modulation rate is, from eqn. 3.7, equivalent to an increase of 3 dB in signal-to-noise ratio. It can be seen that for any error rate greater than about 10^{-6} the uncoded system gives fewer errors than the faster system with coding. Even with more powerful half-rate codes the intersection of the two curves tends to occur at an output error rate of 0.1% or less [1].

The curves of Fig. 11.1 apply for a non-fading signal in white Gaussian noise, under which conditions the distribution of the element errors in the input to the decoding system is random. It is not practicable to apply directly the same method to evaluate the performance of error coding under fading conditions, since under such circumstances the errors will not be randomly distributed but will occur in bursts of relatively high error density when the signal fades to a low level, with a very small error probability during the periods that the signal is near its maximum. The probability of a block being received with more errors than the correcting power of the code is therefore greater on a fading signal than it would be with a non-fading signal giving the same mean error rate.

These two factors (the increase in input element errors caused by redundancy, and the burst pattern of element errors owing to fading) are the primary reasons why the performance of many error coding systems on h.f. is several orders of magnitude poorer than that predicted from basic theory. They also form the basis of the author's opinion, implied above, that the advocacy of error coding techniques as a preferred alternative to the improvement of demodulation techniques is unwise. The use of high-redundancy FEC codes on tactical (i.e. poor-signal, high-urgency) links is particularly inappropriate.

Various so-called 'burst error correcting' codes have been developed in an attempt to overcome the fading problem. However, with the capacity given by modern microprocessor technology to carry out correcting algorithms repeatedly at high speed, a simpler procedure with the same objective may be preferred. Taking as an example the (10, 5, 1) code, 10 such data streams may be time-division multiplexed, the data channel sending one element from each stream in turn (a process known as 'interleaving'). Thus the 10 elements from a single block would be distributed over 100 elements of real time in the signal and therefore the errors in any one burst would be distributed reasonably evenly between the 10 blocks. The burst nature of h.f. errors could in principle be eliminated by such means and the performance improved, *but only to that given by the calculations based on random errors*, at the price of an increase in electronic complexity and decoding time delay. Measurements carried out on channel simulators or on real radio paths show that a very large measure of improvement is obtained for relatively small amount of interleaving but that to approach the random-error level of performance requires a total delay in excess of several seconds (intuitively one would expect a total interleaving delay of at least one fade period to be required).

A minor result of some of these investigations is the confirmation of the intuitive deduction that if there is a simple relationship between the fading rate of the signal and the length of a block in real time, the interleaving process may be less effective. The problem may be reduced by arranging that the time intervals between successive elements of the same block are different, and in a quasi-random pattern (a principle used in some burst-error correcting algorithms).

One factor which should be noted is that the first (and usually the largest) term in eqn. 11.1 is proportional to p^{e+1}. It follows therefore that the longer the code block (inferring the larger the value of e) the greater will be the effect of any factor which causes an increase in the raw element error rate p [1]. It has already been noted that one effect of the redundancy of a code is to reduce the element duration (for constant output data rate) and therefore to increase the raw error rate. It would seem that this effect alone is sufficient to suggest that binary codes using long data blocks with high redundancy would not be effective and that there is an optimum block length for any given maximum acceptable output error rate [2] (as distinct from an optimum total real-time decoding delay in an interleaved system). Although it could be argued that this conclusion contradicts the Shannon theorem (see Section 2.1), the author would suggest that the apparent anomaly may be due to the limitations of binary coding, and that increasing the encoding complexity by coding with a higher modulo arithmetic (see below) may produce an improvement which is not attainable by merely increasing the block length in binary coding.

The complexity of the considerations discussed above suggests that it is difficult to compare effectively different types of code from a purely mathematical standpoint, and probably the clearest guide to the factors involved is in the very limited series of tests carried out on a channel simulator in which different types of FEC code were compared under identical conditions, and in particular the modulation rate of the communication channel was adjusted so that each code gave the same output data rate [4, 5]. The results indicate that where two codes are similar in block length, redundancy and interleaving interval (or the equivalent in a burst error correcting code) they will be likely to produce very similar results, although they may be based on totally different mathematical algorithms. A particularly interesting point is that codes with similar interleaving delays produced results which were apparently a monotonic function of the element period, i.e. a code with a greater redundancy (requiring a higher modulation rate) invariably produced a higher undetected output error rate. Such evidence, and consideration of some of the basic concepts outlined above, leads the author to the opinion that a relatively simple short-block code, heavily interleaved with a random interleaving interval over a large number of time-division-multiplexed channels should give at least an equivalent, if not better, performance than a more complex long-block code giving the same overall real-time decoding delay.

11.3 Binary Coding of Multi-level Channels

Since, as indicated above, the error coding and decoding processes are independent of the demodulation process, it follows that coding may in principle take place at any point in the communication chain between the data source and the point at which signal modulation takes place, and that similarly decoding may take place at any point from the immediate output of the demodulation process to the final data output. An important case in the present context is when information presented in the form of a binary data stream is to be conveyed over a multi-level system, such as one of the MFSK systems described in this book, where it would seem initially that any of the three schemes shown in Fig. 11.2 would be equally possible.

The first alternative, in which coding and decoding is carried out on a binary basis, would seem particularly attractive in view of the fact that most of the earlier published material on error coding systems deals solely with coding for a binary symmetric channel, so that the principles and achievable performance are well established, and many commercial error coding equipments for binary channels are available. In fact it can be simply shown that such a system would be largely ineffective.

The discussion of the creation of errors in an MFSK system by added white noise (Section 3.3) indicates that the tone detector selected in error is equally likely to be any of the other $M-1$†. Considering as an example a 32 tone system where each element is decoded into a 5-element binary sequence, when an error occurs it means that any sequence other than the correct one is equally likely. If a binary code is to be powerful enough to correct any error in a single tone element it must be capable of correcting any pattern of 5 binary-element errors. A (10, 5, 1) binary code would only correct 16% of single-element errors (i.e. those giving a single binary-element error in the output data). *It follows therefore that an error coding system based on binary principles may be comparatively ineffective when used over a multi-level communication link.* Of course this does not imply that the decoding and coding algorithms cannot be carried out using conventional binary logic circuits or microprocessor techniques, but only that the coding arithmetic should be carried out in a modulo related to the number of levels in the communication system *at the point at which element errors are generated.*

11.4 Multi-level Coding – Basic Principles

Before proceeding to discuss this approach, one interesting point should be noted. In binary coding, the redundancy necessary in order to introduce the

†In the case of errors caused by fast fading, Doppler shift or multipath propagation this is not true, there being a bias towards the selection of a tone detector adjacent in frequency to the signal, or that corresponding with the tone element preceding or succeeding the element being detected, but these effects are not consistent enough to affect the underlying assumption.

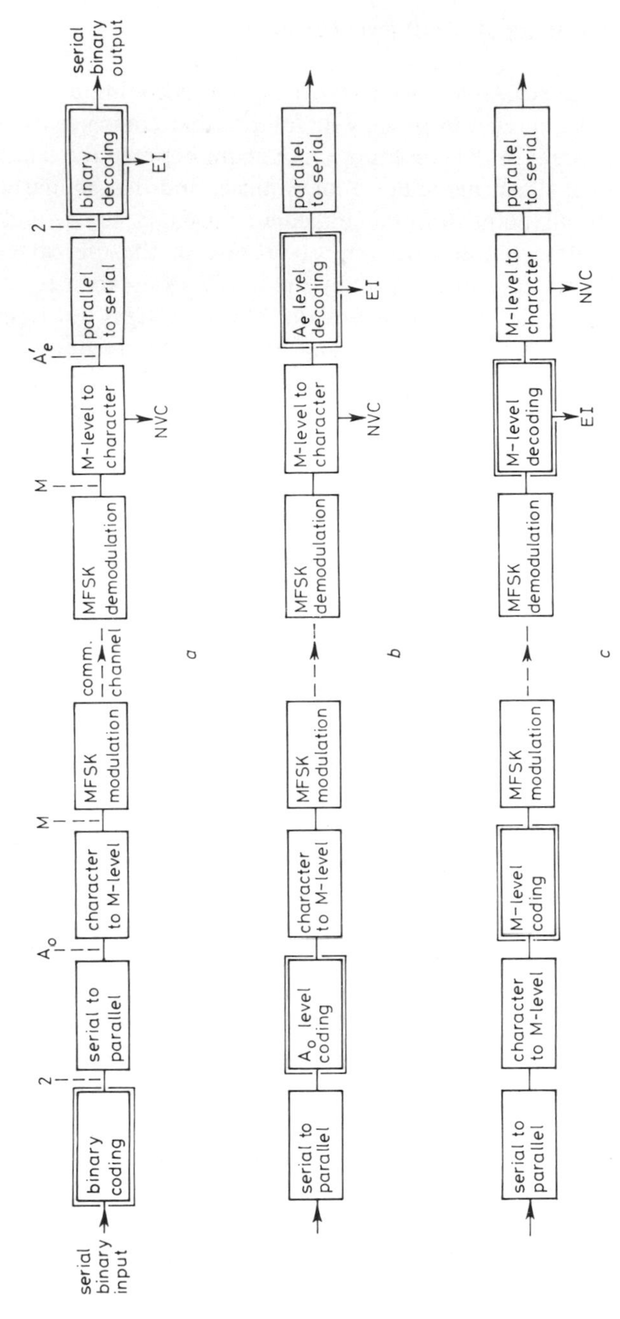

Fig. 11.2 *Alternative methods of error coding*
a Coding at binary level
b Coding at character level
c Coding at *M* level

check information can only be achieved by adding binary elements to the data stream. In a multitone system there are two alternatives, either retaining the same number of tones as for the uncoded data and introducing additional elements for check purposes, or by increasing the number of tones and retaining the same number of elements. The author has investigated both these possibilities but has so far been unable to find any practical method of implementing the second. Even if such a method were to be established mathematically, it would seem a fundamentally inefficient system because of the logarithmic relationship between the information per element and the number of tones. For example, if one check element is added to 9 data elements the redundancy is 10%, whatever the number of tones being used, and the normalised bandwidth is increased by 10%, while with the alternative system, to increase the information per element by 10% for a 32 tone system (say) would require an increase in M to 45. In fact it can be shown that the second process will always require a considerably greater increase in normalised bandwidth than the first. All subsequent discussion will therefore be limited to coding by introducing additional check elements, without modification of the number of signalling levels. Whether the resultant reduction in output data rate is accepted, or whether the element length is reduced to maintain the same data rate and increase the bandwidth is a separate matter discussed later.

Published work on error coding using multi-level arithmetic‡ is still very limited and it is potentially a wide and promising field for investigation. To simplify analysis, attention will be concentrated on the type of MFSK system discussed earlier in this book, and complying with the underlying assumptions and analysis carried out in Chapter 2.

The first stage is to discuss a 'multi-level parity check' technique similar in principle to the binary parity check which forms the basis of many error detection and FEC codes, and for clarity operation at the modulo-M level is assumed. Each tone frequency is allocated a numerical weight of an integer between zero and (M-1) inclusive (or between 1 and M inclusive if this is more convenient). The allocated weighting can be quite arbitrary, for instance it may be convenient for it to be simply related to the frequency, such as giving the lowest frequency a weight of zero, increasing by units up to the highest frequency with a weight of M-1. To form a code block of B tone elements, the weights of (B-1) elements are added, and the sum subtracted from the lowest multiple of M which will give a positive result (the Ms complement). The frequency corresponding to this number is then output as the check element for the block. Thus, taking $M=32$ and $B=8$ as an example, the data stream:

. . . 14/1/7/29/3/6/23 . . .

gives the sum of $83=(3\times 32-13)$ and so would be transmitted as:

. . . 14/1/7/29/3/6/23/13 . . .

‡Note that to be applicable to MFSK any such work must be based on the assumption discussed above, i.e. all errors equally probable. This rules out some published work on 4-level (quadrature) phase modulation.

In the decoding system, the weights of the B characters in a block are added and if the result is not an integral multiple of M (zero, modulo M) then the block contains one or more errors. This process is obviously directly equivalent to an even parity check on a binary sequence.

The exact expression for the probability of an output block containing undetected errors has been shown to be [6]:

$$p_b = \frac{M-1}{M}\left(1-\frac{Mp}{M-1}\right)^B + \frac{1}{M} - (1-p)^B \tag{11.2}$$

Two asymptotic limits for this expression are:

If p is small $\{p \ll 3/(B-2)\}$ then: $$p_b \simeq \frac{B(B-1)p^2}{2(M-1)} \tag{11.3}$$

If the decoding system is fed with random noise, then $p=(M-1)/M$ and:

$$p_b = \frac{1}{M} - \frac{1}{M^B} \tag{11.4}$$

Eqn. 11.2 is plotted in Fig. 11.3 for various values of M and B. The results show an increase in effectiveness with an increase in redundancy and with increase of M, but care should be taken in the literal interpretation of these curves, since the amount of data in a block and the amount of data corrupted by a given number of errors will both vary with the parameters.

An interesting point is brought out by the asymptotic limits of the curves at very high error rates as expressed by eqn. 11.4. If a random binary stream is fed into a simple binary parity check, half the data blocks will be accepted as being 'correct' since it will not detect an even number of errors. If a random stream of modulo-M numbers is fed into a modulo-M parity check, only $(1/M)$ of the blocks will be accepted since errors will be undetected only if the weights 'cancel out'. This simple concept illustrates the considerable superiority of a multi-level parity check over binary at high error densities.

11.5 Optimum Coding Level

The above analysis has described modulo-M coding, assuming that coding takes place at the same level as the number of tone frequencies used. In fact, the multi-level parity check can in principle be carried out in any modulo, but there would seem to be three primary possibilities.

M The number of tones or levels in the communication system.
$A_e = M^C$ The number of potential symbols which can be conveyed by C tone

elements each (which is the 'alphabet' at the output of the demodulation process).

A_o The number of symbols or characters in the original data.

Whichever mode is used, in any comparison care must be taken to ensure that the final results are expressed in terms of the *character* (or byte) error rate of the output data. If the arithmetic is in modulo-M then any calculated output element

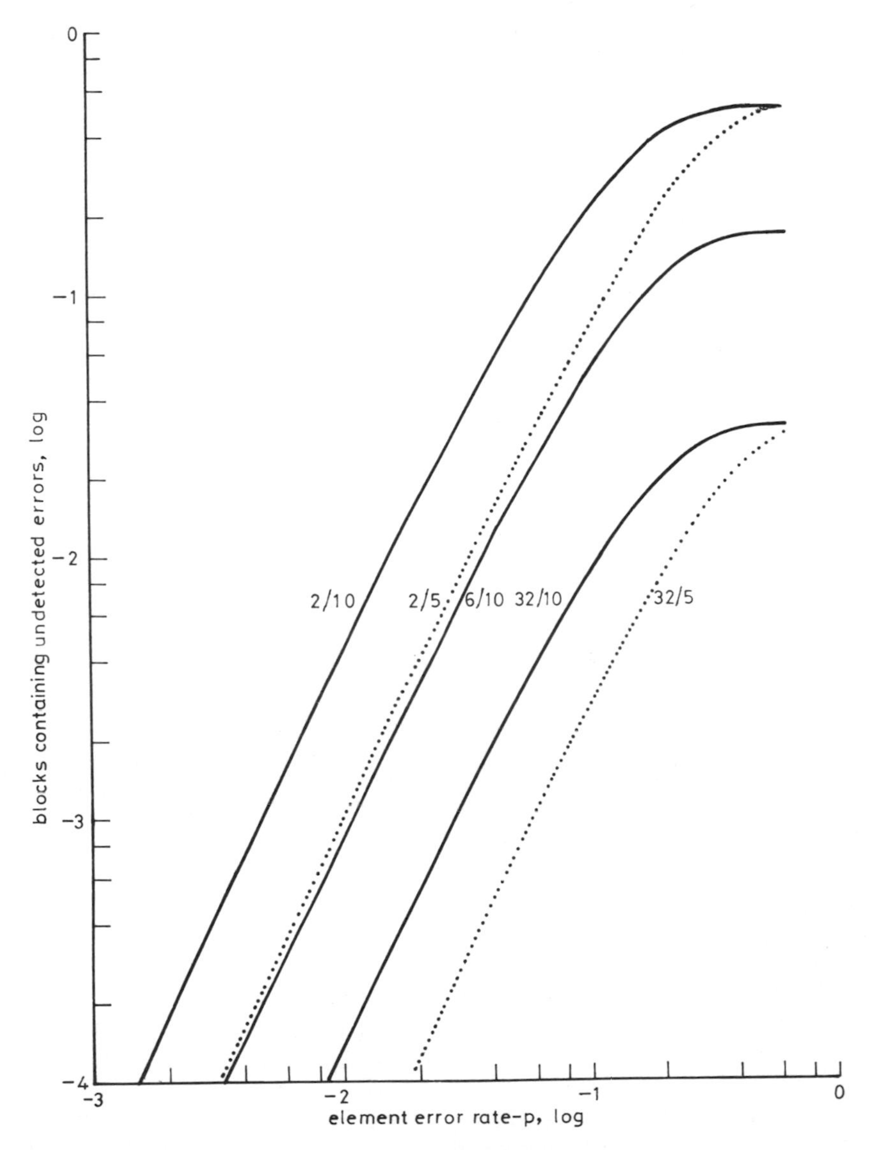

Fig. 11.3 *Performance of M-level parity check*
(Curves labelled M/B)

error rate must be corrected by eqn. 3.16. The correction for the small difference which may exist between A_e and A_o is usually trivial and will be ignored here, it being assumed initially that $A_o = A_e = M^C$. This assumption will enable the terms element, character, frequency etc., to be used in the contexts previously defined and thus minimise possible confusion.

If the modulation system is non-redundant in this sense then any one character at the A_o level will contain the same information as C tone elements at the modulo-M level and the decoding delay caused by a block of B characters will be identical with that caused by a block of (BC) tone elements. One may therefore compare two systems as defined by Fig. 11.2*b* and *c* and Table 11.1, using the asymptotic limits in eqn. 11.3 and 11.4. It can be seen that coding at a higher level A_o reduces the probability of an undetected error block by about $1/M^{C-1}$ *for the same decoding delay and block information content.*

Table 11.1 Alternative multi-level codes

	Fig. 11.3*c*	Fig. 11.3*b*
Error-coding arithmetic modulo	M	$A_o = M^C$
Coding-modulo error rate	p	$1-(1-p)^C \simeq pC$
Block length in characters	B	B
Block length in code elements	BC	B
Prob of undetected error block p_b		
low error rates (eqn. 11.3)	$\frac{BC(BC-1)p^2}{2(M-1)}$	$\frac{B(B-1)(pC)^2}{2(M^C-1)}$
	$\simeq \frac{(BCp)^2}{2M}$	$\simeq \frac{(BCp)^2}{2M^C}$
high error rates (eqn. 11.4)	$\frac{1}{M}$	$\frac{1}{M^C}$

The implication of this analysis is that the multi-level parity check does not depend for its effectiveness on the use of multi-level communication techniques but could equally well be applied to binary transmissions. (It may possibly be shown by more detailed and rigorous analysis that this method is in fact equivalent to one of the known binary coding algorithms). Analysis in this way, however, indicates another factor often overlooked in those theoretical analyses in which the input data is conceived as a continuous binary stream without format, and the calculations are based entirely on a 'bit-error-rate' basis. In fact very many data sources deliver information which is formatted into bytes of constant length and under such conditions it is a very definite advantage if the error code is designed with this factor in mind and processes the data in bytes corresponding to the input data rather than as a continuous stream. The overall conclusion is that *for any communication system* (binary or multi-level), *which transmits information in constant-size bytes, the arithmetic in the coding and decoding algorithms should be carried out in a modulo equal to either the input or output alphabet* (or an integral multiple of this – although this interesting possibility has not been investigated by the author). This means that 7-bit input bytes require a code based on 7-bit arithmetic (or higher).

11.6 Practical Parameters

On the FCO network discussed in Chapter 5, using the Piccolos Marks 2 and 3, the error rate is less than one character in 3000 for most of the time and any form of error correction is unnecessary for the transmission of English text. However, with the growing use of automatic message switching (where a data error may cause a message to be routed to the wrong outlet) and with the increased use of automation and remote control, an error indication code for use in mode (a) of Section 11.2 has been incorporated in the Mk 6 and the work on MFSK coding described here was carried out for this purpose. The extension to FEC and ARQ in later paragraphs has been studied only in mathematical terms and has not yet been implemented.

The first decision concerned the level at which the decoding should take place. Since both the codes used (2 and 5 of Table 6.1) include a certain amount of inherent redundancy there are three separate possible choices of coding modulo for each code:

	ITA–2	ITA–5
Number of tones M	6	12
Input alphabet A_o	32	128
Effective output alphabet A_e	36	144

As shown above, coding at M level is less effective and in this case is also less convenient, since the numbers involved do not lend themselves simply to the binary logic processes inherent in modern i.c. and micrôprocessor technology. The difference in effectiveness of coding between modulo A_o and modulo A_e is small, and coding at the level of the input telegraph alphabets has the considerable advantage of enabling the coding and decoding processes to be carried out in 5-bit or 7-bit binary arithmetic.

A disadvantage of this method is that, since the demodulator system is capable of outputting the larger alphabet, and the decoding system of recognising only the input alphabet, the standby signal and all non-valid characters (see Section 10.2) must be given the same weight as one of the characters of the telegraph alphabet. In this case the character selected was the ‘all space’ character (character 32 in ITA-2 and NUL in ITA-5). Errors in which one character is replaced by another if the same weight (e.g. a standby character being replaced by NUL) are not recognised by the error detection system, but in the circumstances of a practical telegraphy link this is acceptable. The code designed for ITA-5 is therefore based on 7-bit arithmetic (modulo-128), the weight of each character being the binary input (space = 1; LSB first; start, stop and parity elements ignored). When operating in ITA-2 the weights of the last two bits are forced to give effectively 5-bit operation.

The length of error block is primarily a matter of convenience, and in view of the lack of experience and the wide range of possible future applications of the equipment, it was decided to incorporate an alternative choice of either 8 or 16 characters. Eqns. 3.15, 3.16 and 11.2 were then combined to produce the performance curves shown in Fig. 11.4. Laboratory measurements on non-fading signals in white noise confirmed these calculations.

As indicated above, the accuracy of Piccolo is normally adequate for the vast majority of purposes and the very high detection capability of the code at high error densities would suggest that the additional protection given by interleaving may prove unnecessary. It was therefore decided that the initial design should not involve interleaving but that facilities for introducing this by means of an external data processing unit should be included should such measures become necessary at a later date. As implemented (for use in an error indicating mode only) no additional data storage is required in either the coding or decoding circuits.

Of course, the addition of the parity check characters reduces the mean output data rate. It would have been possible to restore the data rate by reducing the element duration which would in turn involve increasing the frequency interval between tones and increasing the bandwidth. Although such a procedure is perfectly practicable for a system in which the error coding is an integral part of the signal processing and the redundancy is fixed, it would have been particularly difficult and inconvenient to achieve in the Mark 6 equipment, which must be capable of operation without EI, or with EI at two different block lengths. It was accepted therefore that the use of error coding would reduce the output data rate proportionally to the redundancy, and so when used in the EI mode some form of automatic control of the date source rate would be necessary (see Chapter 10).

11.7 Forward Error Correction

Although not an immediate requirement, some thought has been given to the possible application of forward error correction coding to an MFSK system such as Piccolo. Such a code, derived from the error indicating code described above, and capable of being implemented by means of a relatively simple control system connected to the existing equipment, has been investigated in principle. In this code one complete block of the error indicating code derived as described above is repeated, giving two separate versions. If either version satisfies the parity check, this is output. If neither version satisfies the check the two blocks are compared, and where two versions of the same character differ, these two are interchanged and both blocks are checked for parity again. If both checks fail

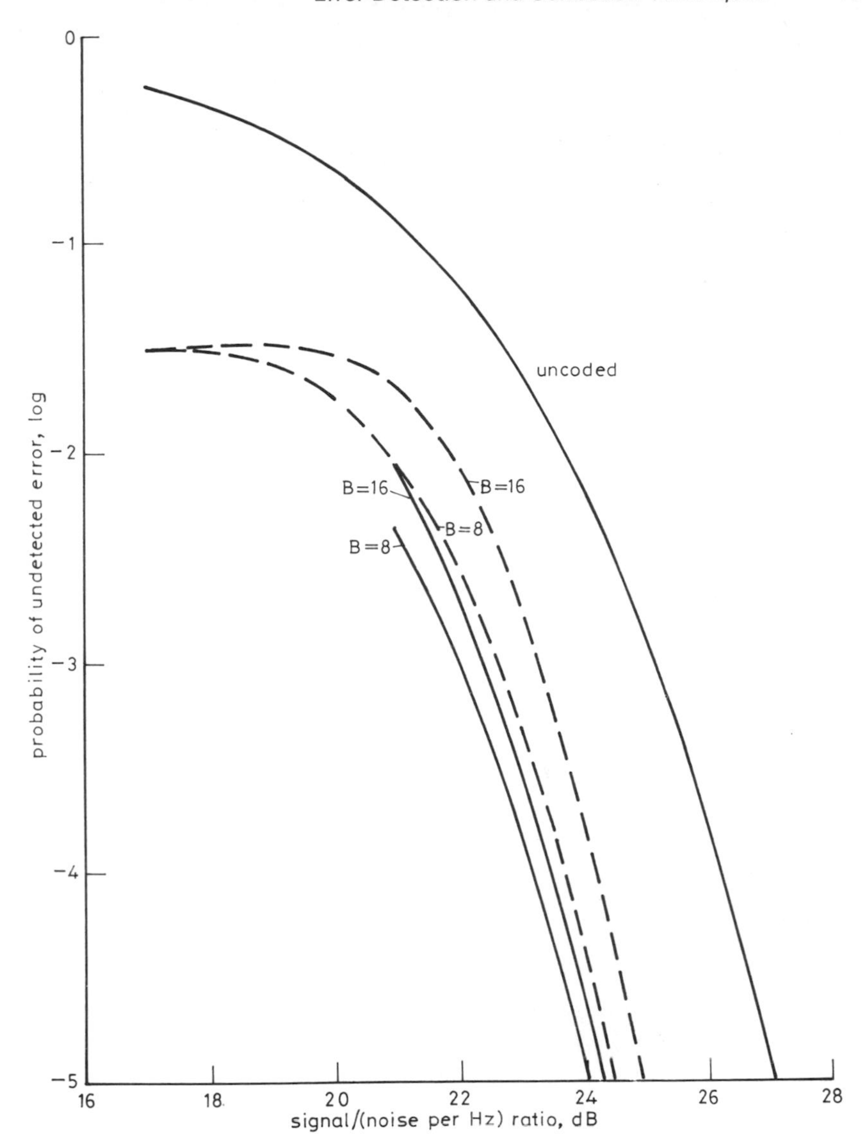

Fig. 11.4 *Performance of Piccolo Mk 6 error detecting code (modulo-32; ITA-2)*
——— undetected character errors
- - - - - - undetected error blocks

again, the process is repeated for all other differences in the blocks until all combinations of differing pairs of characters have been tried. The probability of undetected error of this code is twice that given by eqns. 11.2, 11.3 and 11.4. The probability of errors being detectable but uncorrectable, i.e. that no combination of interchanging can produce a block which satisfies the parity checks, is the probability that an error has occurred on both versions of the same character, which is Bp^2. The rate of such a code is $(B-1)/2B$, redundancy being slightly greater than 50%. However the correcting power is high at high error densities, with most cases of multiple errors being correctable. Forward error correction codes of this type are obviously an alternative to time-diversity combination, and, being entirely digital, are simpler to implement. The relative effectiveness of the two techniques may be studied by the method discussed in the Appendix.

11.8 ARQ (automatic request for repetition)

The use of ARQ systems on binary h.f. links is well established, of which the system recommended by the CCIR [6] is probably the best known. The telegraph alphabet is sent in a synchronous 7-unit format (0.71 rate), using a 3:4 code which will detect all single errors and many multiple errors. The detection of an error character causes the rejection of that character and the two subsequent data characters, these three characters being repeated as a result of the request for retransmission. If any of the three characters are mutilated in the retransmission another request is initiated and so on. Thus under conditions of heavy error rate the mean output data rate may be reduced virtually to zero. Under the conditions for which this system is intended, i.e. multiplexed commercial trunk links, error rates which cause such a large reduction in data rate would not be acceptable, but this effect (coupled with the requirement for a return link) indicates against the use of ARQ on smaller low-power or tactical networks. If required, it is possible to implement an ARQ system on the MFSK link using the error indicating code described above and repeating complete blocks. However, the availability of microprocessor techniques suggests that blocking of the system by indefinite repetitions could be avoided by not rejecting the mutilated block but storing it. The repeated version of the same block, if this too contained errors, would then be compared with the first version as described for the FEC code above. The probability of a second repetition then being required would be negligibly small and there would be no possibility of blocking. Under very heavy error density conditions the total data rate would be somewhat below that of the FEC code (because of the control signals), and the undetected error rate would be the same.

11.9 Block Synchronisation

All systems of block error coding require the decoding system to be run synchronously with the incoming data blocks and therefore a process of block synchronisation must take place before the transmission of data. Since the standby character has been allocated a weight, and provided that this weight is not zero or M, a continuous standby signal can be coded and will produce a known parity character. In the receiving system the complete block pattern consisting of $(B-1)$ standby characters followed by the known parity character is recognised and synchronises the decoding clock (at block rate) to the parity character. Provided that the element synchronising circuit (Chapter 8) is designed so that the periodic insertion of the parity character does not upset the synchronising process, the two processes of element synchronisation and block synchronisation may proceed simultaneously. There is no particular difficulty in meeting this requirement, and in the Piccolo Mark 6 the only effect of the parity coding on the synchronising system is to slightly increase the time required for acquisition of element synchronising.

11.10 References

1 RALPHS, J. D.: 'The limitations of error detection coding at high error rates', *Proc. IEE*, March/April 1971, **118**, (3/4), pp. 409–416
2 BELL, D. A.: 'Limitations of error detection coding at high error rates (Correspondence)', *Proc. IEE*, Dec. 1971, **118**, (12), pp. 1747–48
3 CCIR: 'Automatic error correcting system for telegraph signals transmitted over radio circuits', Kyoto, 1978, **III**, Rec 342-2
4 CCIR: 'Single-channel radio telegraph systems employing forward error correction', Kyoto, 1978, **III**, Report 349-1, pp.139ff
5 CCIR: 'Comparison between forward error correcting systems', June 1969 Doc III/81-E and Doc XIII/135-E, United Kingdom (this is one of the draft documents used to compile Reference 3 in Chapter 11 and contains more detail)
6 VINCENT, C. H.: 'Module-M parity check in MFSK techniques', (Letters) *The Radio and Electronic Engineer*, 1978, **48**, No. 5, p. 248

12

Special MFSK Systems

12.1 System Variations

In Chapter 6 the procedure for the design of an MFSK system against a given set of requirements was discussed with particular reference to a single-channel h.f. telegraphy link at teleprinter speed and intended to operate on poor signals over long-haul routes. It was shown that with the given assumptions as to the propagation phenomena likely to be encountered the requirement can be met by a basic system, although the constraints applied by the propagation effects are rather tight and it is purely by chance that the available range in the time-frequency domain allows reasonably satisfactory communication at the data rate required. Other communication media or requirements may not be so convenient, and it may be found impossible to reach a satisfactory compromise. It is possible to extend MFSK techniques in various ways but it must be accepted that since the performance of the basic system approaches so closely to that of an 'ideal' demodulator for a non-coherent channel, and the practical implementation allows such a close approximation to optimum in most major factors, then any major change in a parameter or principle, introduced in order to meet a special requirement, must almost inevitably involve a compromise which will cause a deterioration in some other respect. Some of the possible alternative options and variations of design are described in this section, and in later sections examples are given of a method of approach in designing telegraphy links for special purposes where such variations may be necessary.

12.1.1 Double frequency spacing

As indicated in Section 4.8, the 'flutter fading' effect occasionally experienced on some transequatorial h.f. links may limit the accuracy of a basic Piccolo system, and there are a number of other potential applications for MFSK signals where rapid fading may be the principal limitation. In cases where bandwidth is not severely restricted, the use of frequency spacing between tones twice that normally used (i.e. a tone spacing of $(2/T)$Hz) gives the improved performance shown in Fig. 12.1 at the expense of doubling the bandwidth. With single-path reception this will reduce the error rate by about 3:1 (Fig. 4.6), but with a dual

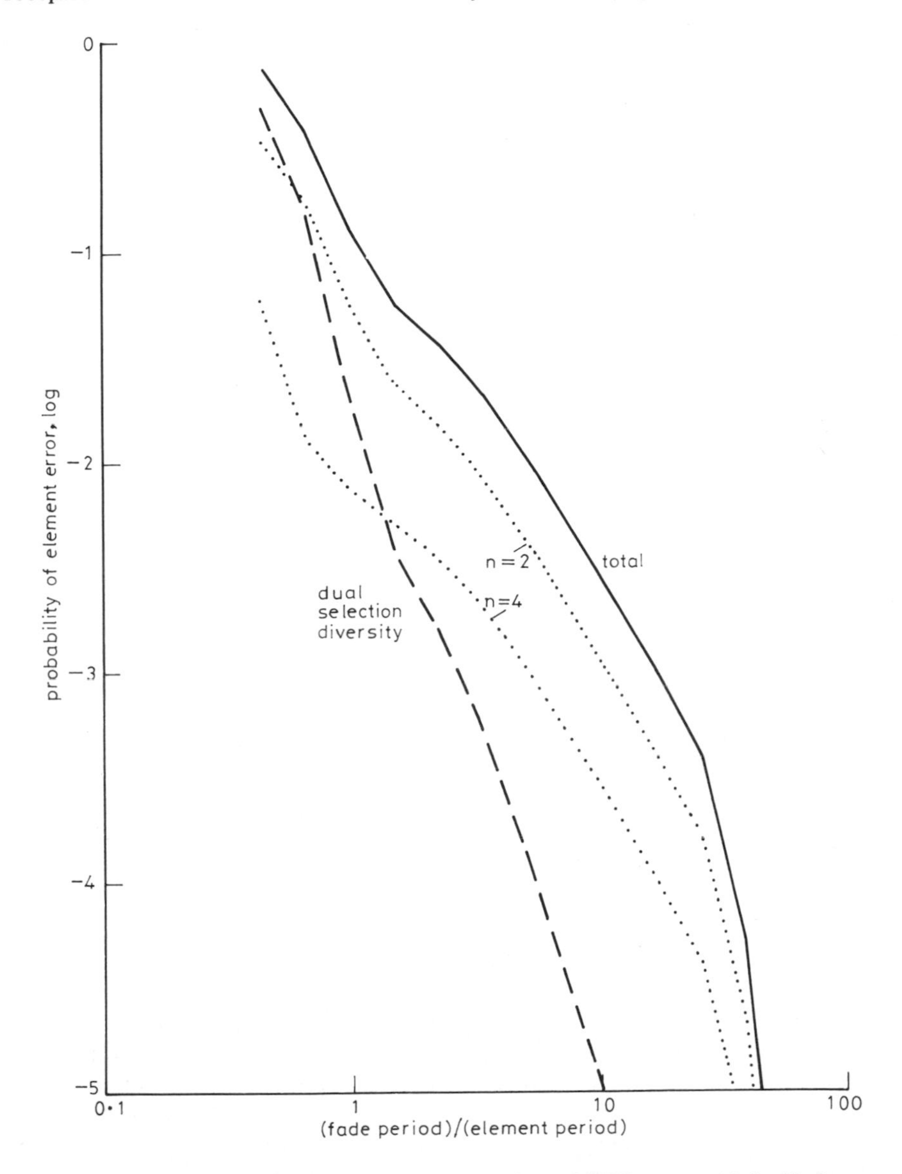

Fig. 12.1 *Theoretical element error rate in fast fading for an MFSK system with double frequency spacing (cf. Figs. 4.6 and 5.3)*

diversity system by approximately 20:1 (Fig. 5.3). However, the curve is extremely steep and even multiple spacing would not (alone) seem to be capable of extending the use of MFSK signalling to cases where the fading rate is appreciably faster than the element rate. In a multiplexed system multiple frequency spacing may be achieved by interleaving the frequencies of different channels, in which case the total bandwidth is not increased.

12.1.2 Signal inversion

A method which has been suggested as a means of combating fast fading is based on the assumption, justified in Section 4.1, that most of the errors caused by fast fading are due to a very rapid change in phase, approaching 180°, occurring when the amplitude of the signal passes through a very low minimum. Calculations based on eqn. 3.5 show that in the extreme case, in which a signal of constant amplitude has its phase suddenly reversed, a selection error will be caused each time the reversal occurs between 0.25 and 0.75 of the element interval. It would therefore seem feasible to attempt to reduce the error rate by a system which continuously monitors the amplitude of the signal, and if this should decrease below a fixed threshold (relative to the median) at any time near to the middle of an element, immediately inverts the signal. Some very rough experiments on these lines have been carried out and showed that some improvement may be possible, but insufficient work has been done to decide whether the complexity is warranted and whether this facility would give a deterioration under some other signalling conditions. A detailed mathematical analysis would be required in order to establish the correct level of the threshold, and this would involve consideration of the relationship between amplitude and instantaneous frequency as discussed in Section 4.1, as modified by any AGC characteristics. The relationships between the time of the null, the depth of the null and the probability of error should also be investigated in greater depth. None the less this idea may be worth developing for links in which fast fading is a severe limitation, and it may be considered as a crude form of adaptive system (see Section 7.4.3).

12.1.3 Doppler shift compensation

The effects of a frequency error between the received signal and the reference frequencies to which the matched filters are sensitive are discussed in Section 3.5. The required stabilities of reference frequencies are discussed in Section 10.4.1 and shown to be within the range of professional standard equipment, and in Chapters 5 and 6 it is shown that the degree of Doppler shift normally encountered in h.f. radio circuits is capable of being accommodated in a relatively simple MFSK system. However, there are cases where this may not be so, and it is then necessary to correct or compensate either the signal frequency or the filter frequencies. If the technique shown in Fig. 9.3 is used to generate the filter frequencies, the correction amounts to no more than moving the common channel (minimum) frequency which will correct all filter frequencies simultaneously, or, if any of the mixing frequencies in the radio receiver can be controlled by a d.c. voltage, the audio signal frequency may be shifted.

Where the frequency error is constant over a long term or may be computed in advance (as would be the case of communication to a space probe or orbiting satellite) this correction may be controlled manually or by computer. In cases where the error is unknown or variable it may be necessary to derive error correction information from the signal itself, and for this purpose it is necessary to transmit a frequency reference signal, either in parallel with the data or inserted into the data stream at regular intervals, depending on the maximum expected rate of change of the frequency error. In the simpler case where the error is changing slowly and is likely to be relatively small, it would be sufficient to insert a burst of the standby signal, and the first phase-locked loop in the synchronising system will generate the required d.c. voltage proportional to the frequency error (see Section 8.7). In cases where the error frequency is changing rapidly it may be necessary to use an additional fast-acting frequency-tracking phase-locked loop in the receiving system, with a sample-and-hold circuit in its control, enabling the loop during reception of the reference signal. This requires a 'block synchronisation' to be established and then for send and receive data rates (which are not affected by Doppler shift) to be maintained throughout the transmission (Section 12.8). Such methods may be electronically complex and the additional redundancy would, of course, increase the normalised bandwidth.

12.1.4 No-repeat coding

The analysis of Chapter 6 suggests that is is quite practicable to select an element period sufficiently long to reduce the errors due to multipath time delays encountered on h.f. propagation to an acceptable level. However, there may be other types of communication medium in which this is not so (an extreme case is discussed in Section 12.6). In such cases there may be a tendency for the energy from one tone element to be sufficiently spread into the subsequent element as to cause errors by the first frequency being selected a second time. The possibility of this occurring may be eliminated by a 'no repeat' code using an additional tone frequency to indicate 'repeat previous element'. Thus, for example, if the 10-tone MFSK telemetry system described in Section 12.3 were required to transmit the sequence:

. . . 3/4/5/5/5/6 . . .

the addition of an eleventh tone, identified as X, would enable the sequence to be transmitted as:

. . . 3/4/5/X/5/6 . . .

In the decoding system, when one element has been decoded, the matched filter selected for that element is disabled during the next element and so cannot be selected in error.

Note that while this technique may be useful in combating extremely long multipath delays, it does nothing to counteract the selective fading caused by multipath, and also that the synchronising system on a link with continuous long-delay multipath must be very seriously considered and may require variations on the techniques indicated in Chapter 8.

12.1.5 'Near Miss' and Soft-decision coding

In the description of the comparator process in Section 9.1, the strobe voltage falls until one of the comparator outputs is energised and this indicates the filter selected by the system. This decision cannot in itself be improved on, but it is possible to provide a measure of the 'confidence' of the decision by allowing the strobe waveform to fall further until it operates a second comparator. The voltage difference between the strobe levels at which the two comparators operate, expressed as a percentage of the amplitude at which the first operates, is an excellent measure of the reliability of the choice. This measurement may be quantified in binary format and used as an additional control variable in an error coding system, the fundamental principle being similar to binary soft-decision coding.

A simplified variation on this idea is incorporated in the Piccolo Mark 6. After the first comparator has operated, the strobe voltage is allowed to fall by a further fixed factor (about 10%). If during this time a second comparator is operated, it implies that the second filter output is more than 90% of the first and therefore there is a measure of doubt as to the correctness of the decision. Under such circumstances the first selection is still output to the decoding system but a control line output from the equipment indicates the occurrence (and also momentarily operates an illuminated display on the front panel). The threshold is set so that the mean number of such indications over a period of time is roughly equal to the number of character errors during the same period. The front panel indication therefore gives an operator an approximate indication of the acceptability or otherwise of the data received, while in an automated system the number of indication pulses output may be analysed to provide a quantified assessment. This indication (referred to as 'near miss') may therefore be used as a control function in a relatively simple ARQ system without error coding, or to provide a selection control in a simple time-diversity system. Such control is, of course, not so effective as that given by a redundant system as described in Chapter 11. A very useful additional facility provided by the near miss indication, however, is that it gives a very rapid indication of the inadequacy or total absence of the received signal, since it operates repeatedly under conditions of severe noise, speech or other quasi-random interference.

12.2 Ionospheric Research

Soon after Piccolo equipment came into use on radio circuits, it was recognised that the type of error observed on the printed copy can often give a general indication of the prevailing conditions on the link, or of some maladjustment of the equipment. More obvious cases are:

(a) Interference from c.w. signals tends to produce repeated blocks of the same character, corresponding to the interfering tone.
(b) Bad synchronising shows as a tendency to repeat the preceding (or following) tone. In a single-element-per-character code this prints double characters.
(c) A serious frequency error, (either from maladjustment or drift of a frequency standard, or from an extreme Doppler shift), causes the selection of the tone detector immediately adjacent in frequency to the signal being sent.

In the Piccolo Marks 1, 2 and 3 the tone frequencies are in the same order as the alphabet and these effects are immediately identifiable, which suggests that MFSK principles (and possibly commercial telegraphy equipments) could well be employed to investigate some ionospheric effects.

For instance, it has already been noted that the flutter fading which occurs on some transequatorial paths is one of the recognisable limitations of MFSK reception and may give an unacceptably high error rate, even when the signal is clearly audible and well above interference and noise. When using the early Piccolo equipment (100 ms elements and 10 Hz spacing) it was noted that on most occasions the errors generated under these conditions were consistently on the same side of the correct tone filter, implying that the spectrum of the signal is not only broadened, but also displaced in frequency by the flutter fading process. No research has been carried out to establish whether this is the result of a true Doppler effect (i.e. a bodily shift of all spectral components by the same amount, maintaining a symmetrical spectrum) or some more complex combination of amplitude and phase modulation giving an assymmetrical skew on the frequency spectrum. It would seem quite feasible to discriminate between these two by a specially designed bank of matched filters spaced at, say, 1 Hz intervals and integrating over one second. Such research could indicate an explanation of the phenomenon, as suggested in Section 4.8.

A matched filter for a continuous tone element of fixed length is not so useful for investigation in the time domain, where autocorrelation techniques may be preferred. However, it is worth noting that the synchronising system described in Chapter 8 is a useful method of establishing the time of arrival of a narrowband signal and with a more sophisticated second loop and preceded by a good audio AGC could be adapted to give a continuous analogue indication of the time of arrival accurate to less than 0.5 ms and with a response time of less than 100 ms.

A particular case which may be of interest occurred during the early trials of the Piccolo Mark 6 system from Australia to the UK. Problems had been encountered in which the synchronising system behaved erratically and on one occasion it was reported that the system was frequently slipping by an element length. Detailed investigation and recording confirmed that this was indeed so and that apparently the incoming signal was occasionally suffering a multipath delay of 50 ms! It was eventually realised that the path delay for a signal to travel from the transmitting station to the receiving station by the shortest Great Circle route was approximately 40 ms but by the longer path on the reciprocal bearing was approximately 90 ms, a difference of almost exactly one element length, and despite the directivity of the transmitting and receiving aerials, propagation conditions at this particular time allowed reception from both directions with roughly equal strengths of signal. Despite the identification of this phenomenon, the author would hesitate to suggest that all h.f. telegraphy systems should be designed to accept values of multipath delay of 50 ms or more!

12.3 Telemetry

This term is taken here to mean the communication of a stream of scientific or engineering data from one point to another. Where these data consist of mixed text and numbers, it is probably simpler to regard it as normal telegraphy using the ITA-2 or ITA-5 code. A particular case, however, is when the data consist of a continuous stream of numerical data alone, where the use of an alphanumeric code is inefficient. This applies to the telemetry system for National Data Buoy DB1, which, as described in Chapter 1, was one of the first applications of MFSK techniques other than the Piccolo equipments.

The DB1 [1, 2] is a large (40 tonnes) unmanned marine data buoy designed initially by the National Institute of Oceanography and funded by the UK Department of Trade and Industry. It was designed to be anchored at ranges up to 300 km from the coast to measure local conditions (meteorological and marine) and telemeter results back to the shore hourly over a low-power simplex h.f. radio link. The telemetry equipment was developed by Racal Electronics Ltd. and used a special MFSK system designed by the FCO (CED). After a year's proving trials moored a few miles offshore in the North Sea, in 1978 the buoy was anchored in the Atlantic on the edge of the continental shelf about 310 km south-west of Land's End, England, and has been operational there continuously to date (January 1982).

The data from the measuring transducers are stored if necessary until required for transmission, then are input to the modulator as a continuous, unformatted stream of decimal digits (actually 3-digit numbers from 000 to 999), all formatting and identification being carried out in the software of the computer at the receiving end. It is therefore essential that character synchronism be maintained throughout the system during the whole of a transmission (see Section 10.3). Two types of data are included, the first half of the transmission consisting of stored data from various meteorological and oceanographic sensors (which require a high standard of data reliability and accuracy), followed by a period of wave-analysis data transmitted in real time at 10 digits per second, but in which lower accuracy is acceptable. Analysis on the basis of Section 2.5 indicated that the simplest MFSK code would probably be as effective as any, i.e. a 10-tone system where each tone represents one decimal digit, leading to a basic system with ten tones (plus synchronising, see below) at 10 Hz intervals and with 100 ms element length. The analysis of error coding in Chapter 11 had not at that time been carried out and so the complete message of stored data is transmitted four times in succession (giving a large measure of time diversity) and decoded with a majority-vote algorithm which also flags each output digit with a confidence rating. It was anticipated that the accuracy without error coding would be sufficient for the real-time data.

Each transmission is preceded by a synchronising period which includes a high-integrity 'start of data' sequence. It would therefore in principle have been possible to have used two of the data tones for synchronising, but in view of the experimental nature of the link and the lack of experience in MFSK techniques at that time it was decided to play for safety and to add two additional tone frequencies at one end of the band for a synchronising signal. Otherwise the element synchronising process is similar to that described in Chapter 8.

Since the initial project study envisaged ranges up to about 600 km, the radio system was designed to use a fixed frequency in either the 4 MHz or 8 MHz band allocated for meteorological use. Most reception would therefore be by ground-wave but additional one-hop propagation giving very deep slow fading and unusually long multipath delays was possible. The signal was likely to be very weak at times, with high atmospheric noise levels. The 100 ms element length approaches optimum for these conditions (see Section 6.5), although the close frequency spacing gave problems at first with stability and microphonics in the transmitter (where environmental problems of shock and vibration are very severe). These were overcome and once the design was established no further problems have been encountered in this area.

In the event, the buoy has operated on a fixed frequency of approximately 4.16 MHz. Field measurements indicate a true radiated power of less than one watt. The aerial is a 10 metre whip mounted on an aluminium mast so that the base is approximately 8 metres above water level. In its latest operation the range to the receiving site is approximately 300 km where reception is by two 10 metre whip aerials in space diversity with a separation of about 100 metres – as much as the site would allow. The report of the performance over the first two years of operation [1] notes that the data recovered was well over 80% and that losses occurred 'more usually from buoy transmission or land line failures rather than through the vagaries of propagation'. Once the system settled down the figures have consistently improved and over the last two years of operation about 95 to 97% of the total transmitted information has been recovered – a notable achievement for a link of this type.

Other telemetry applications are discussed in Sections 12.5 and 12.7.

12.4 Frequency-Division Multiplexing

The previous chapters have discussed MFSK as applied to a single telegraphy channel, but a problem which frequently arises (as discussed in Section 7.7) is to design a telegraphy system which will give a high information density in the fixed bandwidth of a 'speech channel' of 3 kHz. It is standard practice to meet this requirement by the simultaneous modulation of a single transmitter with a number of DPSK or FSK channels in parallel – so-called 'frequency-division multiplexing'. It is quite practicable to employ similar techniques with MFSK modulation and in fact limited facilities for this type of transmission are included in the Piccolo Mark 6 equipment. However, there are a number of specific factors, both advantages and disadvantages, which must be borne in mind.

The first concerns the frequency 'guard band' which is allowed between adjacent channels. With normal FSK this must be sufficient for the channel separation filters to reduce the adjacent channel interference to an acceptable level. However, the orthogonal responses of the matched filter detectors used in MFSK reduce the need for this requirement. If the transmitting modems are arranged to operate synchronously (in the sense that the transitions between tone elements occur at the same time on all channels) and the frequency interval between the end tones of two adjacent channels is made an integral multiple of the matched filter sample period, then signals on one channel are orthogonal to all the detectors in another channel and the need for a large attenuation is very considerably reduced. In the limit the interchannel spacing may be reduced to a single tone interval and channel separation filters eliminated, the full wide-band signal being applied to all tone detectors. (This concept is inherent in the Kineplex and ANDEFT systems discussed in Chapter 7). This arrangement will give the maximum information density and will be assumed in the following discussion.

Then the guard band G (see Section 3.4.4) = 1, so that the bandwidth occupied by a single channel is (M/T) Hz. The number of channels K in a bandwidth B is then given by:

$$K = BT/M \tag{12.1}$$

The total data rate is:

$$H_t = \frac{K \log_2 M}{T} \tag{12.2}$$

giving a normalised bandwidth of:

$$B_o = \frac{B}{H_t} = \frac{M}{\log_2 M} \quad \text{Hz/bit/s} \tag{12.3}$$

The note following eqn. 3.15 suggests that system breakdown occurs when the signal-to-noise power ratio in a single filter reaches a certain critical level, thus for a fixed input noise density one can postulate an 'input signal power per channel at breakdown' W_r defined by:

$$W_r = \frac{r\bar{N}^2}{T} \tag{12.4}$$

where r is a constant factor (suggested approximately 10) times $\bar{N}^2$ the noise power density.

The total received power at this breakdown point is KW_r, so the received signal energy per bit at breakdown:

$$R_r = \frac{KW_r}{H_t} = \frac{r\bar{N}^2}{\log_2 M} \tag{12.5}$$

For a specified error rate, the required ratio of total received signal power to noise density for a fixed bandwidth and total data rate is therefore independent of the number of channels.

However, in practice a telegraphy transmitter is 'voltage limited' rather than 'power limited', i.e. it is the peak input *voltage* which must be held constant as the number of input carriers is varied (see Section 7.7). A more practical criterion is therefore the equivalent 'transmitter rated power' W_t, defined as the power in a single sine wave having the same peak voltage as the composite waveform. Then at breakdown $W_t = K^2 W_r$ and the equivalent transmitter rated energy-per-bit is given by:

$$R_t = \frac{K^2 W_r}{H_t} = \frac{r\bar{N}^2 K}{\text{Log}_2 M} \tag{12.6}$$

The waveform from the sum of several sine waves has a high ratio of peak to RMS value, and some degree of peak limiting may be applied, but insufficient to affect the general conclusions, which are that M should be as large as possible consistent with the required bandwidth occupancy (as for a single channel system), and the number of channels K should be as small as possible. The element length T should therefore be as short as is permissible from multipath and other considerations.

Taking a practical case in which the input data is assumed to be an unformatted binary stream, and since from Fig. 3.5 the normalised bandwidth increases for $M > 10$, it is convenient to take $M = 8$ (i.e. 3 bits per element per channel). If the 3 kHz bandwidth were occupied by a single channel, the required tone spacing would be about $3000/9 = 330$ Hz, giving an element period of 3 ms. This would be impractical on h.f., and the minimum number of channels is therefore determined by the minimum acceptable element length. In most practical cases a multiplex system of this nature is not likely to be used on short-haul links (where cable or microwave would be preferred), and so the considerations on multipath delays in Chapter 6 could be revised towards considering only the shorter delays (say less than 3 ms). One could therefore consider a somewhat shorter element, say about 20 ms, giving 50 Hz tone spacing.

The suggested system would therefore consist of 7 channels, each of 8 tones at 50 Hz spacing. With the minimum interchannel spacing this gives a nominal overall bandwidth of 2.8 kHz. The data rate is 3 bits per element per channel giving a channel data rate of 150 bits/s and a total data rate of 1050 bits/s.

The optimum method of distributing input data of a given format (e.g. ITA-2 or ITA-5 code) between the channels and the optimum error coding algorithms for such a system would depend on the application. Three techniques already in use in FSK signalling could usefully be applied. These are in-band frequency diversity, time diversity, and frequency interleaving (see Section 12.1.1). The derivation of optimum parameters for a given set of practical requirements and involving all these factors constitutes an extremely interesting problem.

The data rate of 1050 bits/s is sufficiently near to the 'standard' requirement of 1200 bits/s as to merit consideration as to how this could be achieved. From eqn. 12.3, $M \leqslant 6$, and two possibilities are suggested:

K	M	T	Frequency spacing	Bandwidth	Data rate
9	6	18 ms	55.5 Hz	3000 Hz	1292 bits/s
7	6	15 ms	66.6 Hz	2800 Hz	1206 bits/s

For any such system it will be necessary to code from a binary input to a non-integral number of bits per element without introducing redundancy. This will cause some deterioration in performance (particularly when an error coding system is used) because the corruption of one signal element can cause an error in more than one character or byte of the output data. The analysis and coding for such systems is therefore considerably more complex than those treated hitherto in this book, although by no means impracticable.

12.5 Satellite Communication and Deep-Space Telemetry

One could propose three possible applications for MFSK techniques in this field:

(a) Intercontinental commercial telegraphy and data communication via large geostationary satellites.
(b) Telemetry to and from small low-power unmanned stations, via either geostationary satellites or more specialised orbiting satellites (similar to the data buoy requirement discussed above).
(c) Telemetry from deep-space probes.

The first possibility may be eliminated on practical grounds alone. Whatever may be the advantages of MFSK techniques, they are extremely unlikely to be applied in such a high-cost commercial area until they have been established in the industry for many years and have become internationally accepted on a large scale. The more conventional binary techniques are so well established that this is unlikely to happen in the foreseeable future. On the other hand it is quite conceivable that after a few years of wider use in h.f. applications, the technique will be sufficiently established and the advantages sufficiently well defined to allow one or more channels of MFSK to be incorporated into one of the smaller and more specialised satellite systems similar to Marisat or Meteosat. Before this can be done, the problems peculiar to this situation would need to be investigated in depth; a subject which is beyond the scope of this book.

The use of MFSK techniques for space probe telemetry is a possibility, at least in principle, and it is useful to consider such a link here as an example of a communication medium in which the primary obstacle to good communication is thermal noise (which is assumed to remain constant as the signal amplitude falls with increasing range). Pure SSB at carrier frequency would almost certainly be impracticable at the radio frequencies used, so some form of sub-modulation would be necessary to reduce the Doppler shift owing to the movement of the probe to a reasonable level. Such Doppler shift as remains (depending upon the type of sub-modulation) would, in principle, be predictable and change slowly and thus could be compensated for on a computerised basis. The available bandwidth would be wide (by normal h.f. standards) but the initial required data rate may be high and a flexible system would be required in which the data rate is progressively reduced as the signal-to-noise ratio deteriorates in order to maintain the optimum compromise between data rate and accuracy. Decoding delay and the capital cost of the receiving modem would not be significant factors.

Assuming that slow long-term frequency drifts can be compensated for in the receiving system, the first stage in the analysis of the problem is to consider all factors affecting the short-term frequency stability and phase fluctuations of the received baseband signal (ignoring noise). From these variations is calculated the maximum usable element length, which determines the orthogonal frequency spacing. The total number of tone frequencies and matched filters is then determined by the available bandwidth.

For the shortest ranges, where the received signal-to-noise ratio is high, the band is divided into a number of frequency-division multiplexed channels of 8 tones each, signalling in 3-bit bytes (as discussed in Section 12.4). The channel frequencies could be interleaved and various diversity and error coding techniques used as previously described. As the range increases, and the error rate rises to an unacceptable level, the transmitting format is changed to 4-bit bytes, using 16 tones per channel and half the number of channels. This will improve the raw error rate in accordance with eqn. 12.6. The process is repeated for longer ranges until each remaining channel is sending 8-bit bytes, using 256 tones per channel. It is probable that the improvement to be gained by further increase is negligible, and for the extreme ranges either higher redundancy correcting codes or higher orders of diversity must be employed (although the latter is of limited value for a non-fading signal).

Although this very rough analysis is based on grossly over-simplified and idealised concepts, there is little doubt that such a system could achieve a marked improvement in performance over its binary equivalent.

In view of the large number of matched filters required, the receiving system would certainly be realised by digital techniques as discussed in Section 9.8.

12.6 Moon-Bounce Communication

As described in Chapter 1, a very early application of MFSK principles was in a digital link signalling by reflection of a 2.6 GHz signal from the surface of the moon [3]. The reference discusses the project in detail and it will only be summarised here as a very interesting example of the adaptation of the techniques for a most unusual case. The characteristics of the path for design purposes are given as:

Doppler shift: varying ±5 kHz between moon rise and moon set, computed in advance to a few tens of hertz.
Fading: very deep fading due to the reflection of the S-band signal from the irregular surface of the moon and the slow 'libration' (apparent rotation over a few degrees) of the moon's surface. The fading rate is 'several fades per second' and although not specifically described as such the fading is presumably Rayleigh in distribution.
Multipath propagation: the largest signal is reflected from the centre of the moon's disc but appreciable power is reflected from the outer edges of the disc (the 'limbs') where the total radio path is increased by almost a moon diameter. The result is that the received signal is a continuum of components of monotonically reducing amplitude at increasing delay, the power density with 0.6 ms delay being about 10 dB down on the central reflection and that at 11 ms delay about 27 dB down.
Signal-to-noise-ratio: measured in an 800 Hz bandwidth was approximately 15–19 dB.
Signalling rate (required): 800 bits/s.

Initial experiments were carried out using binary FSK modulation, received by integrate-and-dump matched filters of the resonant circuit type described in Section 9.1. Five-fold frequency diversity was used with linear addition of the five paths for each frequency. At a modulation rate of 800 Bd the final output error rate was about 20%, at 400 Bd 1.5% and at 200 Bd 0.15%. The binary system was then replaced by a 16-level MFSK system with an element length of 3.33 ms and a dead time (for resonator quenching) of 1.66 ms giving the required overall data rate and an output error rate of 0.3%. Since the major cause of errors was still the very large amount of delayed energy from the previous element, a seventeenth frequency was added to give the 'no-repeat' coding described in Section 12.1.4. The result was a reduction in overall error rate to 0.015% and using this system telegraph traffic consisting of 15 teletype channels at 50 Bd was successfully received.

12.7 Underwater Acoustic Telemetry

As a result of participation by the author in the DB1 project (see Section 12.3) he was asked by Dr J S Rusby, of the Institute of Oceanographic Sciences, to carry out an informal feasibility study on the possible application of MFSK techniques to a short-range ultrasonic telemetry link from a transducer lying on the sea bottom (typically measuring water velocity or temperature) to a tethered data buoy. The model used for the analysis is shown in Fig. 12.2. No attempt was made to examine the fine detail of the system required nor to consider the more advanced techniques developed by specialists in this very difficult field, the primary aim being to estimate the order of magnitude of the following effects, and to consider in general terms possible methods of overcoming them:

Doppler frequency changes due to movement of water in the communication path.
Doppler frequency changes due to vertical movement of the buoy (receive transducer) caused by wave motion.
Multipath propagation caused by reflections from water surface and sea bottom (initially assumed flat and parallel).
Variations in relative time of arrival (synchronisation error) due to water and buoy movements.

The parameters used for modelling purposes were:

c = velocity of sound in sea water	$= 1.5 \times 10^3$ m/s
H = horizontal range	= 200 m
D = depth	= 45 m
f = transmitted signal frequency	= 45 kHz
Horizontal water velocity (tidal stream)	= 3 knots (1.5 m/s)
Horizontal water velocity (surge)	= ± 2 knots (± 1 m/s)
Horizontal surge period	= 5s/surge cycle
Maximum vertical wave height (peak-to-peak)	= 20 metres
Period	= 12 s

Assuming $(h+k) \ll D$ (the optimum situation), it can be shown by simple geometry that the change of transit time caused by the tidal stream is about 0.3 ms, while the periodic variation caused by the vertical movement of the buoy (assumed to follow the wave) is ± 1.5 ms. The Doppler shift in the received frequency from the tidal surge is less than 70 Hz, with a maximum rate of change of about 37 Hz/s. Study of the ray structure indicated in Fig. 12.2 leads to the pattern of received signals indicated diagrammatically in Fig. 12.3, with differential arrival times as tabulated in Table 12.1. Each ray (path) is identified by two parameters (x,n) where x is either s or b, depending on whether the first reflection is from the surface or the bottom, and n is the *total* number of reflections.

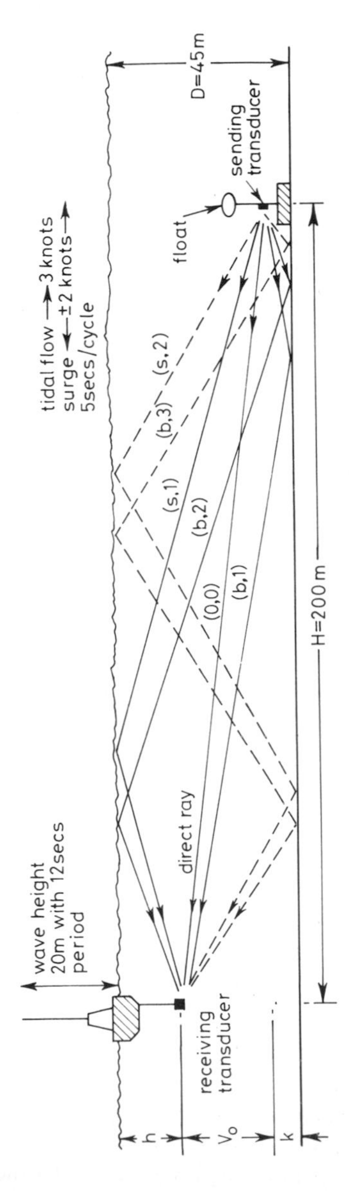

Fig. 12.2 *Model of an underwater acoustic telemetry link*

Table 12.1 *Differential vertical distances and time delays*

	First reflection					
	Surface		*Bottom*		*Differential time Delay*	*Group*
n	even	odd	odd	even	(ms)	
				$2(h+k)-2D$	0	Hypothetical
0	0*				0 (= 136.7 ms)	1st
1		$2h$	$2k$			
				$2(h+k)$		
2	$2D$				24.2	2nd
3		$2h+2D$	$2k+2D$			
				$2(h+k)+2D$		
4	$4D$				64.0	3rd
5		$2h+4D$	$2k+4D$			
				$2(h+k)+4D$		
6	$6D$				112.0	4th
8	$8D$				164	5th
etc.						

*Direct ray (reference). Vertical distance $= D-(h+k)$

If h and k are small, the separation between the first four signals to arrive (the first group) is only a few milliseconds, and they have all travelled very similar paths. After about 24 ms delay the first ray of the second group arrives, following which successive groups are delayed by increasing intervals.

If all reflections were specular and lossless, the inverse square law would apply, so that all rays in the first three or four groups would be attenuated less than 6 dB on the (0,0) ray. From published data [4, 5] the direct ray and those reflected from the bottom only will be fading about ± 10 to 20% on the mean, while reflection from the sea surface will cause deeper fading (typically 40 to 50%), leading to the depth of fading on each ray shown symbolically in Fig. 12.3. Although not described as such, fading will be Ricean in distribution (see Section 4.4) and the degree of signal coherence in each ray separately will be related to the depth of fade, so that any one of the rays in the first group would certainly be adequately coherent for MFSK detection.

It is suggested that the system should be designed to operate on the composite signal from the first group and to ignore as far as possible the succeeding 'echoes', leading to a suggested element duration of 25 ms. If the fading rates of the individual rays are slow compared with the inverse of this (40 Hz), the sum of the four rays should be reasonably coherent over this time, but since destructive interference is possible, the amplitude may be considerably less than the weakest component.

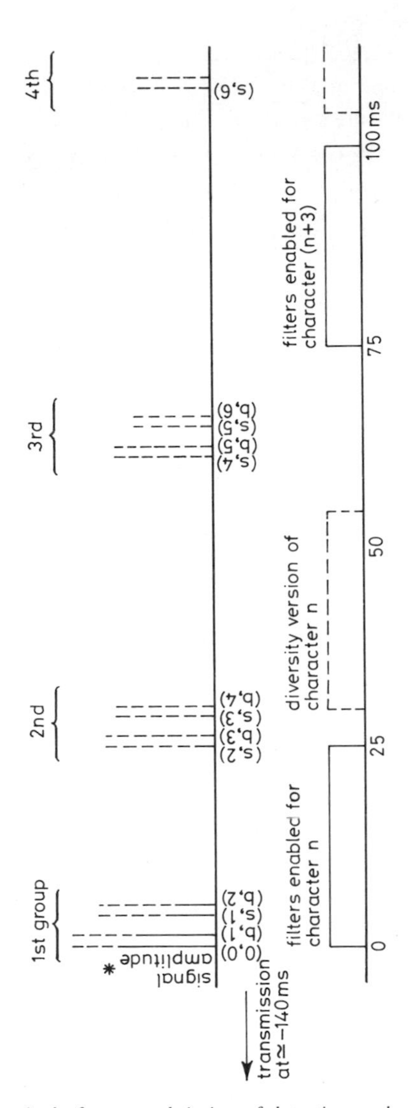

Fig. 12.3 *Times of arrival of rays, and timing of detection cycles*
Time delays between rays in a group are not to scale.
*Amplitudes are indicated symbolically, as are fading depths (dashed lines).

The orthogonal frequency spacing is 40 Hz, but in view of the high and variable Doppler shift, tone spacing at a multiple of this is advisable (see Section 12.1.1). To reduce to a minimum the effects of the later signals a form of 'no-repeat' coding would also seem to be essential (see Section 12.1.4). These two principles may be combined in a scheme such as that illustrated in Table 12.2, which employs three sets of interleaved tones (giving triple the orthogonal spacing) with no-repeat coding applied to each set independently. Thus, during the reception of any one element, the filter at the frequency last received in that set is disabled, as are all the filters in the other two sets.

With this arrangement, it is estimated that the system would be reasonably immune to a flat fading rate of about 40 fades/sec, or a frequency error of about 20 Hz. This is still below the estimated Doppler shift so a frequency-tracking correction system will be necessary, giving a reduction of at least four (see Section 12.1.3).

The basic scheme will convey 40 decimal digits/sec (approximately 160 bits/s) in a bandwidth of 1 kHz. However, the possibility of very low signal levels requires diversity operation. Space and/or frequency diversity should certainly be used as far as possible, but time diversity over a long interval (several seconds or longer) would seem essential. Storage and summation of detected amplitudes as suggested in Fig. 12.4 would be optimum, but if this is impractical, and since error coding will also be necessary, a repetition FEC code (see Section 11.7) should be considered.

An interesting possibility arises from the study of Fig. 12.3, and Table 12.2. The second group of signals which arrives about 25 ms after the first is only slightly attenuated from the main group, and, although fading more deeply, is still not fading 100%. Since these echoes have travelled by a totally different path from the components in the first group the fading should be substantially uncorrelated. There would seem to be no reason why the detectors should not be enabled for a second period of 25 ms during the time of arrival of the second group and so give a form of time diversity due to the differential delay of the different paths without signal redundancy. More accurate and detailed analysis would be necessary to decide whether the timing of the second group is sufficiently accurately known for the second integration period to be timed at a fixed interval after the first, or whether a separate synchronising process is required. Ideally, measurements should be made on an experimental system to ascertain the relative characteristics of the two versions so that if necessary they can be weighted dissimilarly in the combining or soft-decision process to give the optimum output.

The descriptions have been given in terms of a separate filter for each tone frequency, but if the synthesis system of Fig. 9.3 is employed, filter frequencies may be switched to reduce the number required. Alternatively, since a very wide dynamic range would probably not be necessary, it should be possible to realise all signal processing by serial digital techniques (see Section 9.8).

It is possible that a one-shot synchronising system at the beginning of a transmission may not be completely satisfactory because of horizontal movements of the buoy during a transmission as a result of wind and tide changes. It would seem reasonable to combine this problem with that of Doppler correction, and to transmit a separate signal on two additional frequencies, this signal being

Table 12.2 *System with interleaved triple-spaced frequencies and no-repeat coding*

	Tone frequency		
Digit	1st Set	2nd Set	3rd Set
0	300	340	380
1	420	460	500
2	540	580	620
3	660	700	740
4	780	820	860
5	900	940	980
6	1020	1060	1100
7	1140	1180	1220
8	1260	1300	1340
9	1380	1420	1460
X	1500	1540	1580

X = repeat previous digit in same set
e.g. the data: /3/7/2/6/7/6/9 could be sent

/660/1180/620/1020/1540/1100/1500

or

/700/1220/540/1060/1580/1020/1540

or

/740/1140/580/1100/1500/1060/1580

used both for dynamic Doppler correction and for synchronising. The question then arises as to whether this signal should be transmitted continuously in parallel with the data signal (giving an effective reduction of 3 or 6 dB in transmitter power on the data signal – according to the design of the sending transducer and its drive circuit) or whether it should be interspersed with the data at regular intervals, probably between each block of the error code system, so that it also performs the function of providing a block synchronising signal. The optimum of such alternatives is closely tied up with the required data rate, the bandwidth available, and the ratio of received signal to incoherent noise and is therefore outside the bounds of the present analysis, but a rate of change of 40 Hz/s would suggest a maximum refresh interval of not more than 250 ms (10 elements).

Even the shallow and oversimplified analysis given here would suggest that data communication at a reasonable rate over this most difficult medium may prove quite practicable. The principle criticism of the approach described is that the resulting system is optimised for a specific situation, and such developments are often uneconomic. If a number of similar applications are studied, it may be possible to arrive at compromise parameters, or to design a flexible equipment with a choice of parameters. Since the complexity of the post-detection processing would suggest that a microprocessor approach would be essential, the latter alternative would seem favourable.

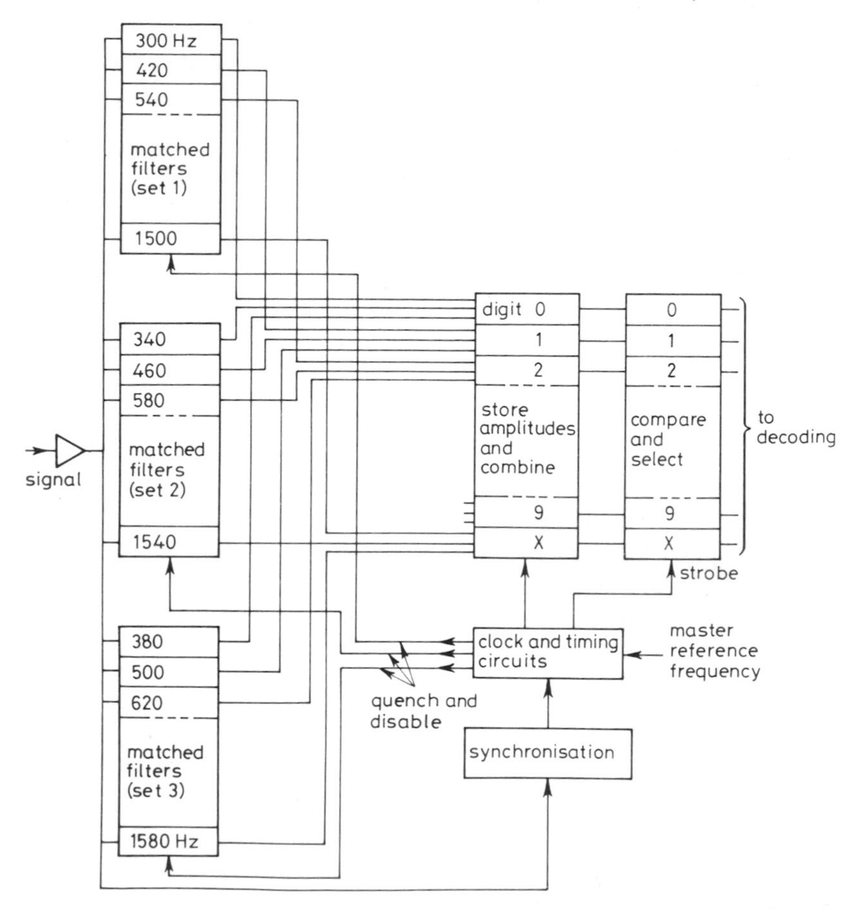

Fig. 12.4 *Block schematic of suggested acoustic telemetry receiving system*

12.8 Ground-to-Air h.f. Telegraphy

At the request of Mr M Maundrell of the Ministry of Defence, the author has given some thought to the possible application of MFSK techniques to the problem of telegraphy from a ground station to a manoeuvring aircraft. As in the previous example, this is not intended to be a serious design study, but rather a consideration of the problems likely to be encountered, and suggestions for an approach to the design. The parameters suggested are therefore tentative, and may be modified as suggested by deeper analysis, experiment or experience.

The model assumed, based on the information provided, is as follows:

Radio link

Ground-to-air, h.f., short and medium range.
Single aerial, single receiver, simplex operation.

SSB reception in 'speech channel' of 3 kHz nominal bandwidth.
Weak signals, with additional nulls due to poor receiving aerial polar diagram.

Aircraft movement

Maximum velocity straight and level: Mach 2 (660 m/s).
Tightest turn (normal flight): 2 nautical miles radius, 240°/min.
Tightest turn (violent manoeuvre): 0.5 nautical miles radius, 18°/sec (communication may be lost temporarily in these conditions).

Data rate required

Continuous telegraphy to subsonic and non-manoeuvring aircraft at 10 characters per second in ITA-2 or ITA-5 code.
Short messages (less than 40 characters) to high-speed manoeuvring aircraft. Total time delay of 5–10 seconds acceptable.

In straight and level flight and receiving the maximum carrier frequency of 30 MHz, the Doppler shift will be approximately 70 Hz, thus a design target of ± 100 Hz maximum allows for an equipment instability of more than 20 Hz. The rate of change of Doppler shift is given by a/λ where a is the acceleration of the aircraft in m/s/s and λ is the wavelength of the carrier (minimum 10 metres). Similar relationships apply for flight in a small circle at a distance from the transmitter, the maximum Doppler shift being given by the velocity of the aircraft and the maximum rate of change of frequency by the radial acceleration (v^2/r). Applying this to the above model, the maximum rate of change in normal flight will be less than 2 Hz/s, and in violent manoeuvres, 10 Hz/s. Optimum codes are 2 and 5 of Table 6.1, but the constraints on the element length, discussed in Chapter 5, are more stringent, and some sacrifice must be made of immunity to multipath in order to gain better immunity to frequency error. Note that unless the aircraft is receiving a strong reflected component from a direction very different from the transmitter, fast flutter fading due to beats between differentially-shifted components is unlikely to be severe.

Fortunately the receiver bandwidth is relatively wide, and since most of the harm done by cross-modulation with interfering signals occurs in the i.f. and detection circuits there seems little point in achieving a very narrow audio bandwidth. It is tentatively suggested that an element length of 30 ms with double frequency spacing (66.6 Hz) is probably about optimum. Since there will be particularly tight demands on the synchronising and Doppler correction system it is suggested that additional tones be allocated for this purpose giving 14 tones in all (in ITA-5). With a total guard band of $4/T$ Hz this gives an occupied bandwidth of $(13 \times 2 + 4) \times 33.3 = 1$ kHz. With a convenient centre frequency of 1.5 kHz the suggested tone allocations are shown in Fig. 12.5*a*. The basic data rate will be 16.6 characters per second, but this will be reduced by the Doppler correction elements and error coding. Variations on this scheme to meet factors not considered may be suggested by the analysis of acoustic unterwater telemetry (Section 12.7).

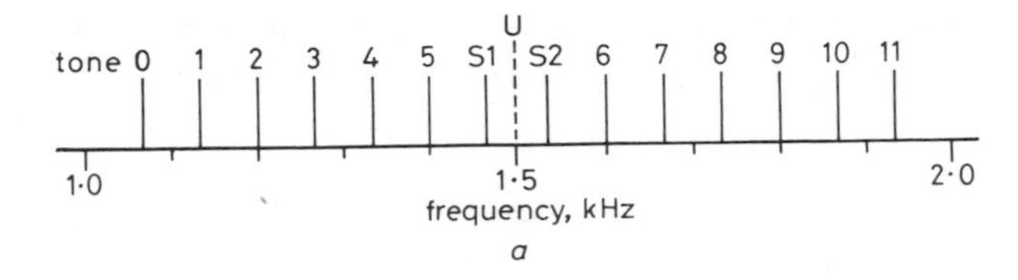

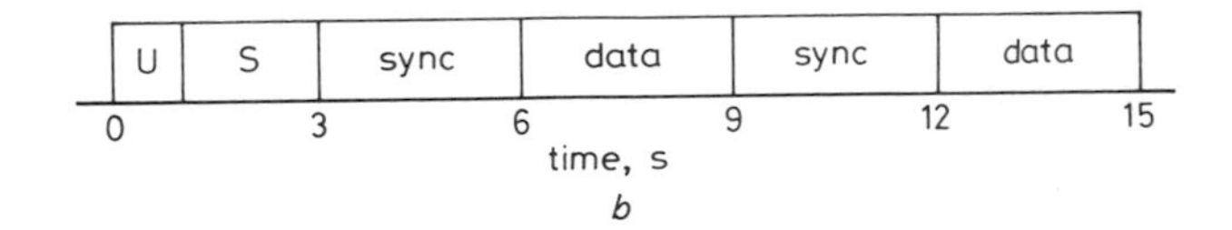

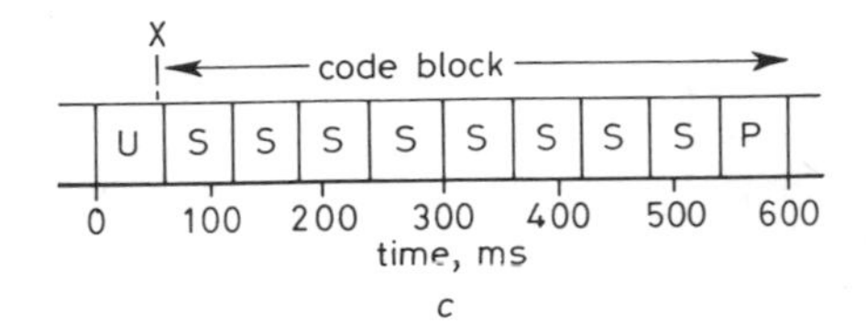

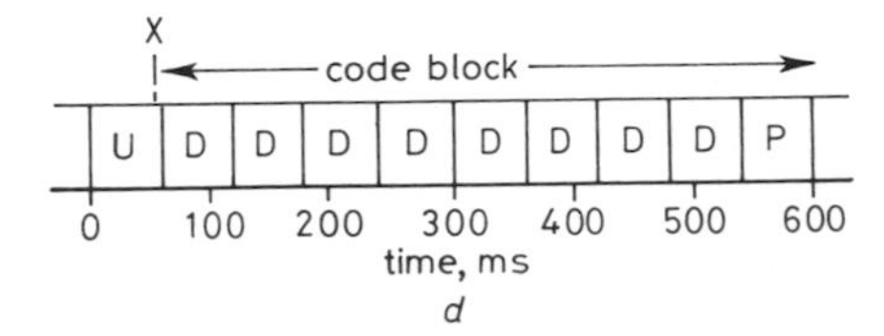

Fig. 12.5 *Frequencies and timing of ground-to-air telegraphy system*
U Unmodulated pilot tone
S Synchronising signal
P Parity check character
D Data character
x Doppler correction sample time
a Arrangement of tone frequencies
b Format of transmissions
c Synchronising block
d Data block

Because of the short operating range (giving possible long multipath delays) and weak signals, in-band frequency diversity is not recommended, and since there are no facilities for space, polarisation or out-of-band frequency diversity, time diversity is indicated. If the additional circuit complexity involved in storage of analogue quantities is not acceptable, then FEC (possibly in conjunction with a soft-decision system as in Section 12.1.5) is the recommended alternative, halving the data rate.

In view of the differing requirements for high-speed manoeuvring and low-speed non-manoeuvring aircraft, it is suggested that an equipment giving at least three different modes is required, all modes being based on a common MFSK modem system and differing only in the data format and the post-detection data processing. The suggested modes are:

(a) Continuous data with interspersed Doppler correction and parity check characters giving a mean data rate of approximately 12 characters per second, with error indication.
(b) As (a) but with a block repeat system after about 20 seconds used either in time-diversity multiplex or FEC code and giving a mean data rate of approximately 7 characters per second.
(c) A short-message mode for messages up to about 40 characters involving a double repetition FEC code but with an initial print-out with error indication on the first transmission, updated and error-corrected by the second. A message of 40 characters fully checked in this way would require a total transmission time of about 15 seconds.

The short-message mode only is discussed here, the other two following by simple adaptation of the techniques illustrated.

The suggested format of a single message transmission is shown in Fig. 12.5*b* and consists of a 15 second sequence as follows:

U = Unmodulated tone at 1.5 kHz to pull in Doppler correction system 1 s
S = Synchronising signal. 1.5 kHz $\pm$33 Hz alternately in *15 ms* elements 2 s

Sync blocks (5 of). Each consists of:
1 character (60 ms) Unmodulated tone as U.
8 characters (480 ms) Sync signal as S.
1 parity character checking the previous 8 Total 3 s
(and for block synchronisation, see Section 11.9)

Data blocks (5 of). Each consists of:
1 character U.
8 characters data.
1 character parity Total 3 s
Sync blocks (5 of) repeated 3 s
Data blocks (5 of) repeated 3 s

The operation of the demodulator system is as described elsewhere and the error indicating code is as described in Section 11.4 with $M=128$, $B=9$. Note that the synchronising waveform is at 33.3 Hz rather than 16.7 in order to maintain a peak phase deviation of 90° and therefore gives element synchronisation only. To obtain character synchronisation the ambiguity may be resolved with the parity check character to the sync blocks, if the tone allocation is made with this in mind.

The major problem is this application lies in the frequency correction and time synchronisation processes, which must include the three operations of Doppler frequency correction, element and character synchronisation, and error-coding block synchronisation, to be carried out in that order. A possible approach is to adapt the synchronisation system of Fig. 8.3 to the more complex arrangements of Fig. 12.6, in which the circuit functions and parameters may be modified by the switches S1, S2 and S3 (shown diagrammatically – probably FET analogue switches or logic elements), to operate in three successive modes:

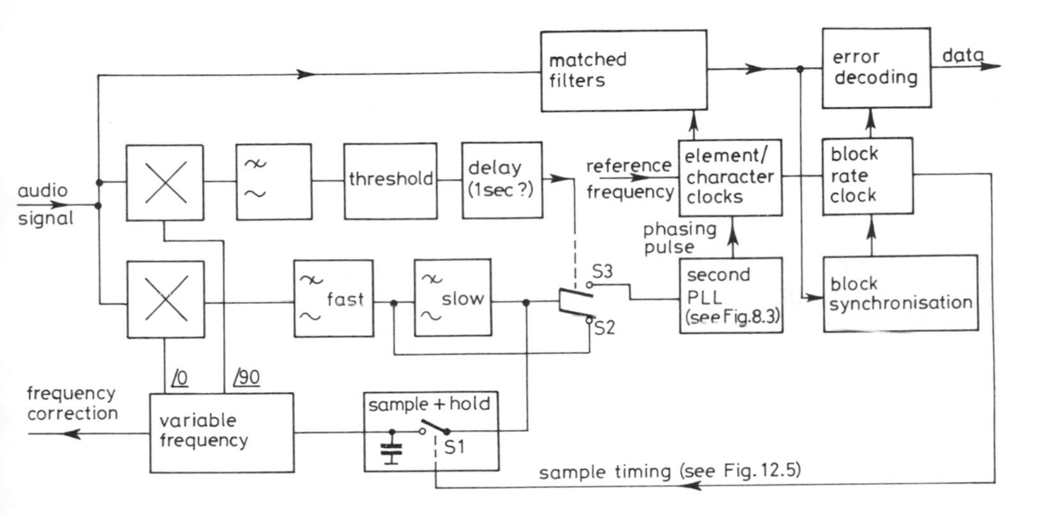

Fig. 12.6 *Synchronising and frequency-correction system*

Acquisition mode: (when no signal is being received): S1 and S2 closed, S3 open. The circuit is 'listening out' for an unmodulated tone and acting as an AFC circuit with a pull-in range of ± 100 Hz and a response time less than 40 ms. The initial unmodulated tone U pulls the loop into lock and generates a steady output from the quadrature detector. After a time delay to confirm this, the circuit is switched to the synchronising mode.

Synchronising mode: S1 and S3 closed, S2 open introducing the 'slow' filter which 'tunes' the loop to 33.3 Hz as described in Section 8.7. When the synchronising blocks S are received, the demodulated signal is passed to the second loop which pulls in to the element timing. After a delay to confirm, a phasing pulse is output to the demodulator timing clock and the next mode energised.

Sample-and-hold mode: S2 and S3 are open, maintaining a slow response. The synchronisation of the code block takes place (operating on the synchronising information as outlined in Section 11.9), following which the switch S1 is closed momentarily at point X of each block, thus 'updating' the frequency correction at 600 ms intervals.

During the data blocks the data is directly printed out (or displayed) with some indication of any block in which an error has been detected. The maximum message length is 40 characters or 5 blocks. At the end of that time another period of synchronisation is transmitted (in case the first attempt has failed completely), followed by the complete repetition of the data blocks. If the first transmission has been unsuccessful these are merely displayed or printed with error indication as before, but if both transmissions are available, they are compared and errors corrected by the FEC code described in Section 11.7.

12.9 References

1 RUSBY, J. S. M., KELLEY, R. F., WALL, J., HUNTER, C.A. and BUTCHER, J. O.: 'The construction and off-shore testing of the UK Data Buoy (DBI Project), Proceedings of the Technical Session (Instrumentation and Communication), Oceanography International 1978
2 RUSBY, J. S. M., and WAITES, S. P.: 'The deployment and operational performance of the DBI Data Buoy', Ocean 80 IEE, Forum on Ocean Engineering 1980
3 WESTON, M. A.: 'Microwave moon communication at high digit rates', *Proc. IEE*, 1968, **115**, (5), pp. 642–651
4 URICK, R. J.: *Principles of Underwater Sound for Engineers* (McGraw-Hill Nov 1967)
5 ALBERS, V. M.: *Underwater Acoustics Handbook, Vol II* (Penn State U.P April 1966) (Chap 5)

APPENDIX

Data rate reduction in a multiplexed system

The discussion in Section 7.7.2 indicates that in-band frequency diversity does not always give the expected improvement in performance, and raises the question, if poor reception conditions are causing an unacceptably high error rate in a multiplexed system, but a lower total data rate is acceptable, what is the best method of sacrificing data rate for accuracy. There are, in fact, at least five alternatives, and assuming that half the data rate is required they are:

halve the modulation rate;
omit from the transmitted signal half of the data tones ('channel shedding');
in-band frequency diversity combination;
time diversity combination;
half-rate forward error correcting code.

The problem will be analysed for the case of a 75 Bd two-phase DPSK system conveying ITA-2 code in a 5 unit synchronous mode, and being received by dual space diversity. The calculations will be simplified by assuming a low element error rate (so that the character error rate can be assumed to be 5 times the element rate) and $R \gg 1$.

Then from eqn. 7.1 the character error rate of the basic system is given by:

$$P_1 = 5p = 2.5/R^2 \qquad \text{(A.1)}$$

If the data rate is halved by reducing the modulation rate to 37.5 Bd, this gives an increase in effective signal-to-noise ratio of 3 dB, so that the character error rate is reduced to $P_2 = P_1/4$.

If the data rate is halved by not transmitting half the tones, then the amplitude of each tone is doubled and the resulting error rate is $P_3 = P_1/16$.

If the overall data rate is halved by dual in-band or time diversity *and assuming uncorrelated fading patterns*, the resulting character error rate is:

$$P_4 = 2.5/R^4 = 0.4P_1^2 \tag{A.2}$$

It is convenient to consider simple half-rate block FEC codes sending one character per block (see Section 11.2). A (10, 5, 1) code is well known (i.e. a code block of 10 elements conveying 5 bits of information and capable of correcting any single element error) and although a (10, 5, 2) code cannot be achieved [1] it is so near to the margin of what is theoretically achievable that it may be conveniently taken as representing an upper bound to codes with similar parameters. For either of these two codes the output *character* error rate is the probability of a block containing undetected or uncorrected errors and first-order approximations are:

$$P_5 = C^{10}{}_2\, p^2 = 45p^2 = 1.8P_1^2 \tag{A.3}$$

$$P_6 = C^{10}{}_3\, p^3 = 120p^3 \simeq P_1^3 \tag{A.4}$$

These equations are plotted in Fig. A.1 and one may draw the general conclusions that an ideal dual diversity system (in addition to the dual space diversity) should give a performance comparable to a simple block code with good interleaving. In-band frequency diversity is likely to be considerably poorer than this for reasons discussed in Section 7.7.2, while time-diversity requires storage of the analogue values of each detector output. What is evident is that although the very simple system of omitting unwanted carriers (channel shedding) is less effective at low error rates, it intersects the curves of the other systems at error levels which could quite easily be met under poor-signal conditions. It is therefore arguable that in many cases it may be preferable to reduce data rate by channel shedding rather than any of the other methods (as assumed for the Piccolo system in Section 7.7.2).

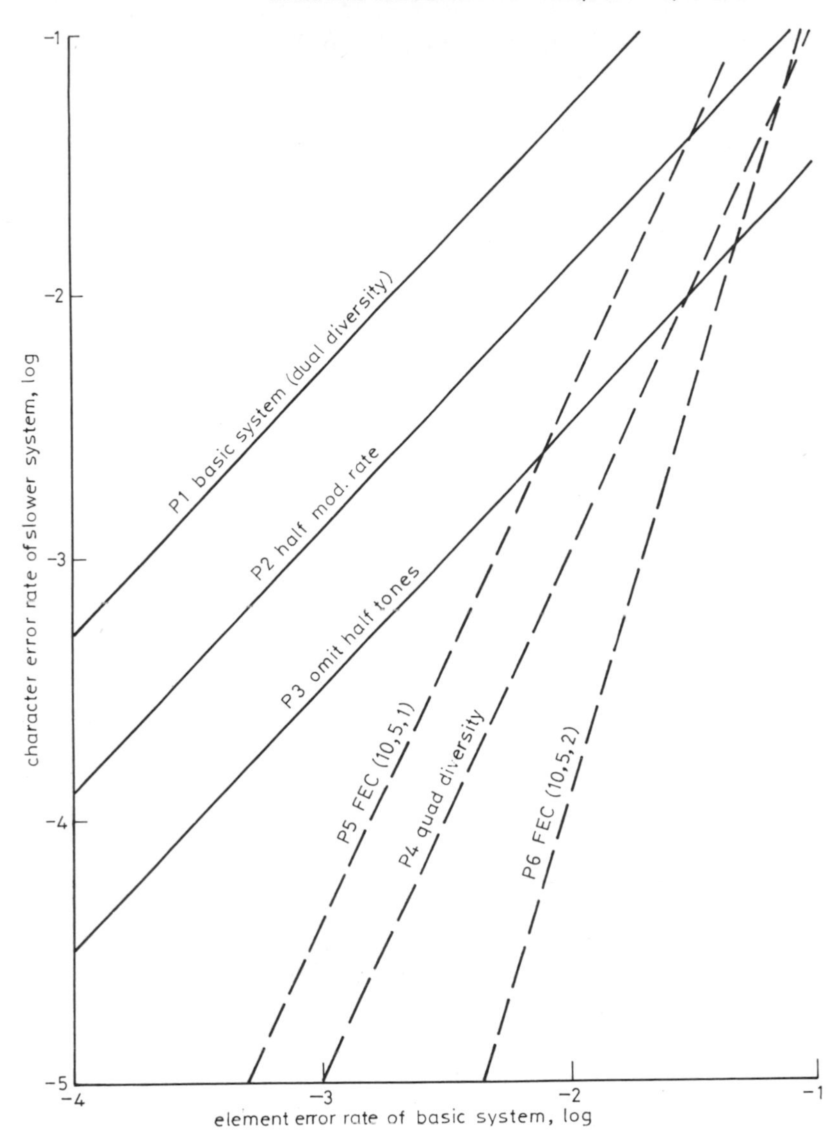

Fig. A.1 *Comparison of different methods of data rate reduction (for binary synchronous 5 bits/character system)*

Reference

1 BELL, D. A.: 'The Piccolo 32-tone telegraph system in diplomatic communications (correspondence)'. *Proc. IEE*, 1973, **120**. (8), p. 851

GLOSSARY OF TERMS AND ABBREVIATIONS

The inclusion of a definition in this list does not imply any general or authorised acceptance, or any significance other than its use in this book.

AGC — Automatic gain control

ARQ — Automatic request for repetition. An error correction system in which the data is coded with redundant information so as to detect certain classes of error pattern. Detection of an error automatically initiates (via a return link) a request for repetition of some or all of the data (Cf. FEC, EI)

bit — Throughout this book the term 'bit' is used as a unit of information defined by the statement: 'A signal consisting of a single choice from M equally probable alternatives conveys Q bits of information, where:

$$Q = \log_2 M\text{'}$$

It is also used to mean a single digit in a number expressed in binary notation. It is *not* used as a unit of signal, i.e. to mean 'a single element of a binary signal' (a common usage which can lead to ambiguity).

BPO — British Post Office (now British Telecom)

byte — The information content (usually in an integral number of bits) of a single symbol or unit of data as input to the system from the data source

CCIR — International Radio Consultative Committee, the technical advisory body to the International Telecommunication Union

CED — Communication Engineering Department of the Foreign and Commonwealth Office of the British Government

character — Most of the discussion is based on the conventional telegraphy alphabets (ITA-2 or ITA-5 *q.v.*) in which each possible symbol is referred to as a 'character' whether or not it prints. Broadly synonymous for most purposes with 'byte'

c.w. — Continuous wave

(D)PSK	(Differential) phase-shift keying. A form of telegraphy in which information is conveyed by digital modulation of the phase of a single tone
DWS	Diplomatic Wireless Service. The organisation responsible at one time for the engineering and operational functions of the FCO communication networks. As a result of reorganisation in 1970–71, it became the Communication Division of the FCO which includes the CED (*q.v.*) and the Communication Operations Department
EDC	Error detection and correction or error detection coding (the abbreviation is avoided in this book because of the possibility of ambiguity). See 'error coding'
EI	Error indication. An error control system in which the data is coded with redundant information so as to detect certain classes of error pattern. The occurrence of such an event is indicated to the receiving operator or equipment for appropriate action to be taken (Cf. ARQ)
element	(Signal element). As applied to FSK, MFSK and PSK systems, the minimum duration of time for which the transmitted signal frequency, phase and amplitude remain unchanged (a constant for the system), or the signal transmitted for one such interval.
error coding	A general term covering all forms of error control by redundant information, including EI, FEC, and ARQ
FCO	Foreign and Commonwealth Office (of the British Government)
FEC	Forward error correction. An error control system in which the data is coded with redundant information so as to enable some classes of error pattern to be corrected (without the use of a return link)
FSK	Frequency shift keying. In this book the term is used in a general sense to apply to any form of telegraphy in which binary information is conveyed by a sequence of tone elements at either of two frequencies. As such it includes systems variously described as frequency exchange keying, two-tone keying, voice frequency telegraphy etc., and detection by filter-assessor or limiter-discriminator techniques (see Section 7.5)

h.f.	High frequency (radio). The band 3–30 MHz
i.c.	Integrated circuit
i.f.	Intermediate frequency (of a radio receiver)
ITA-2 and ITA-5	The standard alphanumeric codes for radio telegraphy, used here in a general sense, irrespective of the actual format of transmission (i.e. whether start–stop or synchronous etc.). For present purposes ITA-2 is identical with CCIT-2, Murray, Teletype 7-unit etc. and ITA-5 with CCIT-5 and ASCII
ITU	International Telecommunication Union
LC	Inductance-capacity (resonant circuit of series or parallel connection of these)
LSB	Least significant bit; lower side-band
matched filter	For the general definition of the term, reference should be made to a textbook of communication theory. It is used here in the sense of an optimum method of detecting a single tone in noise, as defined in Section 2.3
MFSK	A form of multiple-frequency telegraphy in which digital information is conveyed by a sequence of tone elements selected from a set (more than two) of orthogonally-related frequencies, detection being achieved by matched-filter techniques or the equivalent.
Piccolo	The various MFSK telegraphy modems designed by and manufactured for the UK Foreign and Commonwealth Office Communications Engineering Department were given the code name 'Piccolo' (because of the sounds made by the Mark 1). The name is becoming a generic term for an MFSK system but this is to be deprecated
PLL	Phase-locked loop
PSK	See DPSK
PTD	Path time delay
RMS	Root mean square

SSB Single side-band

tone Most of the discussion is in terms of signal elements each consisting of a burst of audio tone of constant amplitude. However, there is no reason why the frequencies used should be restricted to the audio band and the use of the term is retained even when referring to a radio-frequency signal

USB Upper sideband

GLOSSARY OF SYMBOLS

This list does not include terms used in one section only, or those always defined in the text in the paragraph in which they are used.

A Signal amplitude (general)

A_n Amplitude of nth component

A_o Amplitude of component for $n=0$ (usually d.c. or 'wanted' signal)

A_o Input alphabet (Chapters 2, 3 and 11) = number of different symbols or characters which may be input from the data source

A_e System alphabet = number of different symbols capable of being recognised by the demodulator system

B Bandwidth in Hz

B_o Normalised bandwidth or bandwidth occupancy $= \frac{\text{bandwith}}{\text{data rate}}$ Hz/bit/s.

C Capacity (Chap 9, Section 3.1, Figs. 2.2 and 2.3 only, see also LC in Glossary of Terms)

C Number of signal elements used to convey one byte or character

H Information rate in bits/s (suffixes as appropriate)

M Number of matched filters (tones, levels) in system

N Noise voltage (general)

$\bar{N}$ RMS noise voltage in a bandwidth of 1 Hz

$\bar{N}^2$ Noise power density, noise power per Hz bandwidth (note that in many cases, unity signal power is assumed, and then $\bar{N}=1/R$)

Q Quantity of information per byte or character (bits)

R Amplitude of a radius vector (Chapter 4)

R Ratio of (signal power)/(noise power per Hz bandwidth)

R_o Normalised signal-to-noise power ratio = $\dfrac{\text{signal energy per bit}}{\text{noise power per Hz bandwidth}}$

T Sampling or integration period of a matched filter (secs)

T_e Duration of a signalling element (secs)
(After Chapter 3 it is usually assumed that $T_e=T$)

V Voltage (general)

V Amplitude of the envelope of the output of matched filter at time T

V_o V for the 'correct' filter, i.e. corresponding to the signal frequency

V_n V for filter n (defined as required)

Index